T0342663

NATIVE BIRDS OF AOTEAROA

NATIVE BIRDS OF AOTEAROA

Michael Szabo

With an introduction by Alan Tennyson

Illustrations by Pippa Keel

CONTENTS

INTRODUCTION

Aotearoa New Zealand is famous for its birdlife. While the diversity is not large, many species are well known because they are so unusual.

The country's geographical isolation has led the bird fauna down an interesting evolutionary path. The lack of land mammals (apart from bats) has allowed birds to dominate niches occupied by these animals elsewhere. It has also meant that the birds could hide from their main predators – other birds. Birds are mostly visual hunters, unlike mammals, which rely on smell to catch their prey. Camouflage and being nocturnal are common ways birds hide from avian predators, so these are common features of species found here, meaning that few are brightly coloured. Kiwi are famous examples of birds with these traits.

These environmental forces have also favoured the evolution of many flightless species of birds. In a temperate climate like that of Aotearoa, where there is a reliable food supply year-round, there is no need for birds to move long distances (and use up a lot of energy) to find enough food, and hiding, rather than flying away, was an effective way to avoid predators. Sedentary island birds also often evolve into larger forms compared with their ancestors. Without the need to fly and with their size making it more difficult to do so, some birds evolved flightlessness. Some, like takahē (page 39) and kākāpō (page 95), became the largest of their kinds in the world.

However, flightlessness did not always go hand-in-hand with gigantism. The snipe rail, Hutton's rail and four kinds of wren, all now extinct, were small flightless birds. Some birds that became very large were in fact very good flyers: the kererū (page 31) is an unusually large fruit-eating pigeon, the extinct Haast's eagle (*Aquila moorei*) was the largest known eagle in the world, and the extinct Eyle's harrier (*Circus teauteensis*) was the largest known harrier in the world.

Unfortunately, flightlessness left the birds of Aotearoa very vulnerable to the changes brought about by humans. At least twenty of the large flightless species went extinct following human arrival because they were a rich food source and easy to hunt. Moa (all ten kinds) are the most famous of these extinctions, but giant geese (*Cnemiornis*), which weighed up to 18kg, and predatory adzebills (*Aptornis*), up to 19kg, were others. The carnivorous mammals that accompanied people had similarly dramatic impacts, and these remain the primary threat to bird species today. Rats, cats, mustelids, pigs and dogs contributed to the extinction of at least twenty-five kinds of bird. Many other large flightless species, including the South Island takahē (page 39) and kākāpō (page 95), have suffered drastic range declines and have required ongoing human intervention to save them.

Seabirds are another defining feature of the Aotearoa fauna. More kinds of albatrosses and penguins breed here than anywhere else, and the same is true of the less appreciated shags and petrels. One kind of petrel – the tītī (page 73, also known as sooty shearwater or muttonbird) – is one of the most abundant seabirds in the world. Its nesting colonies dominate many islands in the south of the country, where there is an annual harvest of chicks by Rakiura Stewart Island Māori. Included in the seabird fauna are the largest marine birds: royal albatrosses (toroa) have three metre wingspans and breed on southern offshore islands, as well as at the mainland nesting site at Pukekura Taiaroa Head. In all, 173 kinds of seabirds have been reported from Aotearoa, ninety-nine of which are regular breeders and the other seventy-four are either regular migrants or strays. With an archipelago stretching from Rangitāhua Kermadec Islands in the north to Motu Ihupuku Campbell Island in the south, the country covers 2500km of latitude, providing access to vast areas of the southern oceans where seabirds can feed.

Seabirds generally have far less defined ranges than land birds, which therefore have been more vulnerable to land-based human impacts. Only a few seabirds have been lost, but some still teeter on the brink of extinction, including the tara iti, New Zealand fairy tern (page 59). In addition, seabirds face a range of threats at sea that have proven difficult to manage, particularly fisheries by-catch, pollution and changes in sea temperature. Today, most seabirds are confined to nesting on islands free of introduced predators because their habit of nesting on the ground makes their eggs and chicks vulnerable to predation. People lucky enough to have visited New Zealand's remote offshore islands, such as the subantarctic islands, marvel at the numbers and diversity of seabirds present.

The recent fauna of Aotearoa includes 588 kinds of birds (counting species and subspecies, based on the 2022 *Checklist of the Birds of New Zealand*), with 211 of these being endemic (that is, they breed only here). The total continues to grow because new colonists establish every few years and new strays are detected most years. Sadly, the living fauna is just a shadow of its pre-human glory, with the most unusual kinds predominant among those lost to extinction. Sixty-nine birds, including sixty-two land and freshwater birds and seven seabirds, are extinct. If the strays, birds introduced by people, post-human colonists and regular migrants (totalling 225) are removed from the list, then the picture is sobering, with about a fifth of the original fauna having gone extinct.

Birds, being arguably the best-loved and most-studied group of animals on Earth, are very well known compared with many other organisms. Nevertheless, even the Aotearoa bird fauna continues to reveal discoveries. In the last ten years alone, the southern Cook's petrel, Whenua Hou diving petrel, Foveaux shag and Otago shag have been recognised as being distinctive living endemic birds, while the New Zealand swan, Chatham Island swan, Chatham Island merganser, Chatham Island crested penguin, Richdale's penguin, Imber's petrel, kohatu shag, and the Chatham Island kākā have been recognised as extinct taxa.

Many Aotearoa bird species have long histories on the now largely sunken continent of Zealandia. Those with the earliest pedigrees are the extinct moa, kiwi and the New Zealand wrens (including tītitipounamu, page 103; and pīwauwau, page 105), which have been present on Zealandia for most of its existence (it split from other remnants of Gondwana between 83 million and 52 million years ago). Even so, the ancestors of these birds may have flown to Zealandia across the newly formed Tasman Sea. The ancestors of every other kind of land and freshwater bird almost certainly arrived over water, although many of them did so several million years ago, including the large parrots, the honeyeaters, the wattlebirds, hihi (page 117) and the Mohouidae/New Zealand creepers. Others have much shorter ancestries, with close relatives overseas (e.g. ruru, page 89; riroriro, page 111; and pīwakawaka, page 123), and some common birds have colonised the country from Australia only in human times, including kakīānau black swan (page 21), pūkeko (page 41) and tauhou (page 133). Birds that colonised in human times might be more appropriately termed 'native invaders' because they are only present in Aotearoa as the result of human modification of the environment and some have negative impacts on 'true native' species. Finally, many species make Aotearoa their home for only part of the year, including cuckoos and many Arctic waders (e.g. kuaka godwits, page 53) as well as some seabirds (e.g. tītī, page 73). Ancient fossils reveal that, while modern species may be relatively recent arrivals, some of their more distant relatives were present an extraordinarily long time ago. This is best exemplified by penguins, which have a rich fossil record in Zealandia dating back more than 60 million years.

The unique bird fauna of Aotearoa is globally renowned but it has suffered terrible losses through extinction and range contractions. An enormous ongoing effort is needed to protect what is left.

BIRD COLLECTIONS AT TE PAPA

The Museum of New Zealand Te Papa Tongarewa houses the most comprehensive and important collection of Aotearoa birds in the world. It also holds significant collections of birds from other places, particularly South Pacific islands. There are about 80,000 bird specimens preserved in a variety of ways: about 50,000 are fossil bones, 16,000 are 'study skins' (taxidermied lying down), 7000 are skeletons, 3000 are eggs, 2000 are mounts (taxidermied in a life-like pose), 900 are preserved in alcohol, 800 are tissue samples (primarily for DNA) and sixty are nests. The collection is constantly being added to through donations and fieldwork.

Understanding the whakapapa (origins and history) of birds, and their fascinating lives, leads to greater appreciation of te taiao (the natural world) and our place in it. Te Papa's collections form the heart of the museum. As kaitiaki (guardians) of these taonga (treasures), caring for them is at the core of our work.

Te Papa began as the Colonial Museum in 1865, so the collection has been building for more than 150 years. In the initial decades of the museum, before birds had legal protection, specimens were usually collected by being shot, and numerous specimens were then exchanged overseas for foreign specimens. These transactions were poorly documented because specimens were not registered, and so they cannot be easily traced today. The modern focus is on collecting locally occurring species and ensuring that all specimens are registered and looked after in climate-controlled storage conditions, with strict pest control. With most native species now having full legal protection, specimens are obtained through donations of birds found dead, mainly by researchers, Department of Conservation staff and the public.

The specimens are used for a wide variety of purposes. Only a small number are suitable for display and the majority are used for research. They form a fundamental basis for understanding current and past biodiversity. Many specimens of the same species are needed in order to examine variation due to age, sex and moult. A particular role of museums is to hold and care for 'name-bearing' types (those to which the original scientific name is applied) and examples of extinct species. Museums hold valuable series of specimens collected over time, which can be analysed for changes, including past distributions. New technologies also influence the kinds of collections kept today – for example, tissue samples are now used widely for molecular techniques, such as analysis of DNA and isotopes.

Scientific and mātauranga Māori research is conducted by both Te Papa staff and external researchers, such as professionals and students, who either request information or visit the collections themselves. Research outputs help to make the collections accessible and are numerous, including peer-reviewed and popular scientific articles, books, media stories, blogs, social media posts and exhibitions. Artists visit to illustrate specimens and historians examine the inter-relationships between science and history. Increasingly, Te Papa's collections are being made more widely available through the Collections Online website (collections.tepapa.govt.nz). Loans are made to other institutions for research and exhibitions. In-person and virtual behind-the-scenes tours are provided to schools and other interested groups.

The multitude of recent research outputs based on Te Papa's collections includes descriptions of the Campbell Island snipe, two recently extinct close relatives of the hoiho (page 67) and three new species of kiwi (one living and two fossil forms). Other discoveries have revealed that Aotearoa once had two kinds of prehistoric swan that have since been 'replaced' by the kakīānau black swan (page 21), and that kea (page 99, now confined to Te Waipounamu South Island) once lived in Te Ika-a-Māui North Island. But Te Papa's research is not limited to working on its collections. The museum's scientists carry out fieldwork throughout the country – for example, surveys led by Te Papa have discovered and documented many new tītī (page 73) colonies in Ata Whenua Fiordland. Te Papa staff were also part of the team that discovered the only known breeding site of the takahikare-raro New Zealand storm petrel on Hauturu-o-toi Little Barrier Island, a species that was thought to be extinct for more than one hundred years until its rediscovery in 2003.

Te Papa's relationships with iwi, private landowners and community groups (such as Birds New Zealand Te Kāhui Mātai Manu o Aotearoa) are fundamental and much of our work could not be done without their support. Through this broad community, we can inspire awareness and help birds face the ongoing pressures from human changes. Numerous groups, particularly Forest & Bird and Birds New Zealand, already lead and support conservation initiatives to protect and enhance native bird populations. The Department of Conservation leads most threatened species management, and government-led initiatives, such as Predator Free NZ, are inspiring a new generation of conservationists to dream about how our native birds might be able to recover and thrive in the future.

Alan Tennyson, Curator Vertebrates, Te Papa

ABOUT THIS BOOK

This book is inspired by *Native Birds* (1948) and *More Birds* (1951), from the popular Nature in New Zealand series published by AH & AW Reed. Alexander Wyclif (Clif) Reed was a prolific author and published these books under the pen name Charles Masefield. According to *The House of Reed 1907–1983: Great days in New Zealand publishing* (2005), by Edmund Bohan, Reed was able to use printing blocks he had bought of native flora and fauna illustrations long out of print.

The beauty and appeal of these books lies in their simplicity. We have retained a similar approach and even reused the original illustrations where possible. Not every species we wanted to include in this book had original illustrations, and, despite their charm, some were not as accurate as required. Therefore, a number have been carefully created or redrawn in the same style, with the guidance of the authors, by illustrator Pippa Keel.

We have selected sixty birds that represent all groups found in Aotearoa – forest, garden, wetland, coastal, alpine and marine birds – reflecting the range of subtropical, temperate and subantarctic habitats found in Aotearoa. Readers will be able to use the book to identify many common birds, but also learn about some that are lesser known but equally significant in our avifauna.

The birds are grouped following the convention of the 2022 edition of the *Checklist of the Birds of New Zealand*: kiwi, swans and ducks, grebes, pigeons, cuckoos, rails, wading birds, gulls and terns, tropicbirds, penguins, albatrosses, shearwaters, gannets, shags, herons, hawks, owls, kingfishers, falcons, parrots, New Zealand wrens, honeyeaters, warblers, wattlebirds, hihi (page 117), New Zealand creepers, fantails, Australasian robins, grassbirds, swallows, white-eyes, and pipits.

THE BIRDS

KIWI-NUI
NORTH ISLAND BROWN KIWI

Apteryx mantelli

Kiwi are unusual in having various land mammal characteristics. They are flightless and have shaggy hair-like feathers, cat-like whiskers, nostrils at the tip of the bill and large, visible ear openings. The kiwi-nui is the only kiwi species found in the wild in Te Ika-a-Māui North Island. Like all five kiwi species, it is nocturnal, so is more often heard than seen. Large (40cm long; 2–2.7kg), wingless and tailless, it has dark brown feathers, streaked reddish brown and black, a long pale bill, short dark legs, large feet with sharp claws, and tiny vestigial wing stumps. Paired mates call occasionally each night to advertise territory and maintain contact with each other. The male makes a repeated high-pitched ascending whistle, and the larger female responds with a deeper, throaty cry.

Habitat: Native forest and scrub, pine forests and rough farmland from sea level to 1400m, north of Te Āpiti Manawatū Gorge. More widespread and abundant prior to human colonisation, the species now has a fragmented distribution and declining populations.

Ornithologist's notes: During the day, kiwi-nui rest in burrows or under thick vegetation, emerging after dark. They use their sharp claws to fend off competitors and predators. Pairs are monogamous and build a nest in a burrow at the base of a hollow tree or log. Egg-laying can take place in any month, with the peak season from June to November. The female lays 1–2 huge white eggs, which are incubated for 75–90 days by the male. Replacement egg-laying occurs in November–January. The chicks become independent at 2–10 weeks old. In common with all kiwi, the birds have weak eyesight but very good senses of smell, hearing and touch. They feed by slowly probing the ground with their long bill for small invertebrates (especially worms and larvae), and sometimes catch freshwater crayfish from streams.

Status in Aotearoa: Endemic

Conservation status: Not threatened

RORORA

GREAT SPOTTED KIWI

Apteryx haastii

The roroa is the second-largest of the five kiwi species (45cm long; 2.2–3kg). Nocturnal, flightless and tailless, with tiny vestigial wing stumps, it has shaggy brownish-grey plumage that is finely mottled with white. It has a long, narrow cream-coloured bill with nostrils at the tip and sensitive whiskers around the base, short grey legs, and large feet with sharp claws. Females are larger and longer-billed than males. At night, males make a high-pitched ascending whistle, and females make a slower and lower-pitched ascending trill. Pairs call occasionally at night to advertise their territory and maintain contact with each other. Duetting pairs alternate their calls.

Habitat: Native forest, scrub, upland tussock and subalpine zones of north-western Te Waipounamu South Island. Now confined to north-west Nelson, the Paparoa Range, and Kā Tiritiri o te Moana Southern Alps between Te Kopi o Kaitangata Lewis Pass and Te Ara Kuiti o Tarapuhi Arthur's Pass. A new population has been established in Nelson Lakes National Park, and supplementary birds have been released in Wharepapa Arthur Range and Nina Valley.

Ornithologist's notes: During the day roroa rest in a burrow, rock crevice or hollow tree, emerging shortly after dark. Monogamous and fiercely territorial, they use their sharp claws to fend off competitors and predators. Females lay a single very large white or pale green egg from July to December. Incubation is shared by the pair, with males doing most of the incubation during the day and the duties being shared through the night. When they detect prey they probe their bill into the leaf litter or a rotten log, occasionally plunging it deep into the ground. They eat small worms and larvae, centipedes, beetles, wētā, snails, freshwater crayfish, and small fallen fruits.

Status in Aotearoa: Endemic

Conservation status: Nationally vulnerable

KAKĪĀNAU
BLACK SWAN

Cygnus atratus

Although present when Māori arrived in Aotearoa New Zealand, kakīānau were no longer living here at the time of Pākehā colonisation. They were reintroduced in small numbers from Australia initially in the 1860s, but their distribution and abundance within a few years suggests that natural recolonisation may have occurred at about the same time. The kakīānau has a long neck, red bill, red eyes and dark legs, and is usually seen gliding gracefully on water. Reaching up to 6kg in weight and 140cm in length, the swans are ungainly on land and in flight. The sexes look alike, but the female is smaller. The young (cygnets) are grey with a black bill. The main call is a descending squeaky honk or musical bugling. In flight, the white flight feathers make a whistling sound.

Habitat: Resident on lakes, wetlands, farm/oxidation ponds and tidal estuaries.

Ornithologist's notes: Males raise their crinkled upper wing feathers in courtship and threat displays. Both mates engage in elaborate greeting displays with head-flicking and hiss threateningly when defending themselves or their cygnets. Most breeding is by monogamous territorial pairs defending all or part of a pond or lake. Nesting starts in July on a large mound of vegetation at the water's edge. Females lay six large, pale green eggs and incubate them for about a month, mostly on their own. The cygnets fledge after four months, but families may still remain together for several months. Colonial nesting generally occurs in September–November. Herbivorous, they breed only on fresh waters.

Status in Aotearoa: Native

Conservation status: Not threatened

PŪTANGITANGI
PARADISE SHELDUCK

Tadorna variegata

This colourful goose-like duck, with distinct male and female plumages, is New Zealand's most widely distributed waterfowl. The female has a striking white head and light chestnut plumage, contrasting with the male, which is almost uniformly dark grey-black with a dark green sheen on the head. The bill and legs are dark grey-black. Both sexes typically measure 63–70cm in length and weigh 1.4–1.7kg. The male makes a loud goose-like honk in flight or when alarmed, while the main female call is higher in pitch, and more rapid and persistent during flight.

Habitat: Occurs on the three main islands and all large nearshore islands, from the high country to lowlands on grassland in pastoral landscapes, large lakes and urban parks. In the 1800s the species was restricted to the south and east of Te Waipounamu South Island, but the development of farm pasture with exotic grass and ponds has since caused a spectacular increase in its range and numbers.

Ornithologist's notes: The birds form territorial monogamous pairs to breed, occupying a territory for most of the year that usually contains a waterbody used by the ducklings. Females prospect for nest sites in tree cavities up to 25m above the ground, in rock crevices or in holes in the ground. Egg-laying usually takes place in August–September. Clutches of 5–15 large white eggs have been reported, but most clutches of more than twelve eggs are from two females. The female incubates the eggs for a month, and both parents guard the ducklings until they fly. Some broods may join together. A family can remain together for several weeks before the parents leave to join communal moulting sites. Pūtangitangi graze on leaves, seeds and herbs and some terrestrial and aquatic invertebrates.

Status in Aotearoa: Endemic

Conservation status: Not threatened

WHIO
BLUE DUCK

Hymenolaimus malacorhynchos

The whio is the iconic 'torrent duck' of New Zealand's fast-flowing mountain rivers. It is a large (55cm long; 770–900g) blue-grey duck with a dark chestnut breast and yellow eyes, and a pale pink bill with a rubbery black flange that protects it from abrasion when feeding. Males are larger, with more chestnut on the breast and more greenish iridescence on the head. They give a high-pitched, wheezy whistle, *whi-o*, hence the species' Māori name. Whio also have a blunt bony spur at the joint on each wing, which the males use to spar with during territorial fights.

Habitat: Formerly inhabited lakes and rivers from high altitudes to areas of bush-clad lowland rivers and lakes, but now with a patchy distribution on rivers in forested headwaters along the predominant mountain ranges of both main islands.

Ornithologist's notes: One of ten native duck species that breed in Aotearoa. Its blue-grey plumage provides perfect camouflage against river boulders. Birds are generally seen year-round as monogamous pairs that are fiercely territorial, with both sexes defending a 1km stretch of river. Nesting starts in August–September. The nest is hidden in a riverside cave or hole, or close to the river at the base of ferns or a fallen tree. On average, the female lays six large, pale white eggs. She then incubates them while the male waits nearby. Both parents guard the ducklings until they fledge after ten weeks. The ducklings have large feet when they hatch, allowing them to swim in fast currents. Whio eat freshwater invertebrates such as caddis-fly larvae, which they hoover up from the riverbed in shallow water while facing upstream.

Status in Aotearoa: Endemic

Conservation status: Nationally vulnerable

PĀPANGO
NEW ZEALAND SCAUP

Aythya novaeseelandiae

This highly sociable diving duck is common around Aotearoa. The population was estimated at 20,000 in the 1990s, but a 2020 survey found it was more likely to be closer to 5,000–10,000. Compact (40cm long; 600–799g) and blackish, the pāpango has a profile like a toy rubber duck. It spends much of its time underwater, where it can swim a surprising distance, and floats with cork-like buoyancy. Males are dark black-brown with an iridescent blue-green head and wings, paler mottling on the chest and underparts, yellow eyes and a blue-grey bill. Females are dull chocolate brown with paler underparts, brown eyes and a grey bill with a white patch at the base. The male has a high-pitched whistling call, *weeee weo-weo weo-weoooo*. The female utters a low *wack-wack*.

Habitat: Patchily distributed on freshwater dune lakes in Te Tai Tokerau Northland and Manawatū; inland lakes in Waikato, Taupō, Rotorua and Te Matau-a-Māui Hawke's Bay; Te Tai o Poutini West Coast lakes; north Waitaha Canterbury waterways; and eastern and southern high-country lakes.

Ornithologist's notes: A gregarious duck that forms dense rafts in autumn and winter. Courting males engage in aggressive displays. The nest is a tidy bowl made of vegetation lined with a layer of down, and is well hidden on the ground close to water. Breeding takes place mainly between October and March. Monogamous pairs nest alone or in a loose colony. The female incubates 2–13 creamy-white eggs and cares for the young, with the male often nearby. Young ducklings often form crèches. The species has more than twenty Māori names and once had the alternative English name of black teal. Pāpango dive for snails and larvae.

Status in Aotearoa: Endemic

Conservation status: Not threatened

WEWEIA
NEW ZEALAND DABCHICK

Poliocephalus rufopectus

This small (30cm long; 230–270g) aquatic diving grebe uses its large, powerful lobed feet to propel and steer itself in the water. The adults are brown-black with a long neck, a dark head streaked with fine silvery feathers, yellow eyes and a short black 'dagger' bill. In breeding plumage, the front is rich rufous brown. The tail is a very short tuft of black silky feathers. Both sexes are alike. Chicks hatch with irregular black-and-white-striped markings, a pale bill and a small patch of red facial skin. Juveniles retain the stripes until their adult plumage develops, and the bill turns black. Adults give short, chattering calls during the breeding season and when alarmed. The species has a submarine-like ability to change its buoyancy by adjusting the angle at which it holds its dense waterproof feathers against the body.

Habitat: Small lakes and sheltered inlets on larger lakes, from Te Tai Tokerau Northland to the top of Te Waipounamu South Island. Most live in the Central Plateau, Rotorua Lakes area, Te Tai Tokerau, Te Matau-a-Māui Hawke's Bay and Wairarapa. Weweia declined in the South Island in the 1800s–1900s until going extinct there as a breeding species in 1941, with introduced predators a likely factor in its decline. Since 2012, multiple pairs have bred in the Whakatū Nelson and Te Tauihu Marlborough regions.

Ornithologist's notes: One of four native grebe species that breed in Aotearoa. The birds are monogamous and breed year-round, and are aggressive towards intruders. Territorial displays begin in June–July, with an elaborate courtship featuring ritualised displays. Solitary pairs nest on lakes. Females lay 2–3 bluish eggs in August–March. Incubation lasts three weeks and chick-rearing three months, with both parents assisting. Chicks ride on the parents' backs for the first few weeks. The birds eat insects, small fish, crayfish and tadpoles.

Status in Aotearoa: Endemic

Conservation status: Recovering

KERERŪ
NEW ZEALAND PIGEON

Hemiphaga novaeseelandiae

Known as kūkū and kūkūpa in Te Tai Tokerau Northland, this large forest pigeon (50cm long; 630g) is familiar to most New Zealanders. It has a red bill, eyes and feet. The head, upper breast and wings are iridescent green, and the neck, back and upper wings have a purple-bronze sheen. The white of the breast and belly is sharply delineated from the green upper breast. Mostly silent, kererū make occasional quiet, throaty *oo* calls, or brief louder *oo* calls when alarmed. During the breeding season both sexes fly up high over the forest before making a steep skydive.

Habitat: Widespread in large areas of native forest, exotic plantations with an understorey of native trees or shrubs, and rural and urban habitats. The species' populations declined after human arrival due to the use of fire to clear land. After Pākehā colonisation, people hunted the birds with guns and used large saws or axes to cut down native forests for timber. The species has since been protected, so hunting is now illegal. Annual New Zealand Garden Bird Survey counts show a 102 percent increase in kererū numbers over the period 2012–2021.

Ornithologist's notes: The smaller of two native pigeon species in Aotearoa, the other being the parea (Chatham Island pigeon, *Hemiphaga chathamensis*). Both are among the largest forest pigeon species in the world. Kererū are monogamous and breed during September–April. Females lay a single large white egg on a nest platform made of dead twigs. They incubate it from late afternoon to mid-morning, after which the male takes over. The egg hatches after a month. Both parents produce a protein-rich 'crop milk' to feed the young chick. Fledging is about a month later. The birds eat forest fruits, flowers, buds and leaves from a variety of native and exotic plant species, making the species an important disperser of native forest seeds.

Status in Aotearoa: Endemic

Conservation status: Not threatened

PĪPĪWHARAUROA
SHINING CUCKOO

Chrysococcyx lucidus lucidus

This small (16cm long; 23g) annual spring and summer migrant to Aotearoa from Papua New Guinea and the Solomon Islands has a distinctive whistling call that announces the arrival of spring. Pīpīwharauroa are iridescent emerald green above and white below with dark green horizontal bands, and have a short, slightly downcurved bill. They are well camouflaged and inconspicuous, which allows them to conceal themselves in dense foliage. Their main call is a loud, upwardly slurred whistle, rapidly repeated several times. The sequence usually ends with one or two downwardly slurred whistles.

Habitat: Present in Aotearoa in spring, summer and autumn only, in or near forest and scrub, and in farmed and urban areas – reflecting the wide distribution of the species' primary host, the riroriro (page 111).

Ornithologist's notes: Pīpīwharauroa gather in small feeding groups after arriving and then display communally high in a tree. Being brood parasites, they first push one of the riroriro host's eggs out of the nest, before laying one of their own olive-brown eggs in its place. Females lay eggs mostly in November, then the adult cuckoos take no further part in breeding. The cuckoo chick ejects all the other eggs and/or chicks from the nest and is dependent on its foster parents for food for several weeks after fledging. Pīpīwharauroa can eat toxic insects such as hairy caterpillars that are avoided by many other birds – they squeeze out the insides, swallowing them and then discarding the hairy skin. The birds gather in feeding flocks in February–April before departing on their transoceanic migration back to Melanesia. This is the smaller of the two migratory cuckoo species that breed in Aotearoa, the other being the koekoeā (page 35).

Status in Aotearoa: Native

Conservation status: Not threatened

KOEKOEĀ
LONG-TAILED CUCKOO

Eudynamys taitensis

The enigmatic koekoeā is a large (40cm long; 125g), slender annual spring and summer migrant to Aotearoa from Polynesia, Melanesia and Micronesia. The species' annual transoceanic return migration of up to 6500km each way is the longest over water of any forest bird. The koekoeā is dark brown above with black horizontal barring, and buff below with brown and black streaks. It also has brown and black barring across its buff face, and a pale, slightly decurved bill. Its long white-tipped tail makes it look rather falcon-like in flight and it is sometimes mobbed by smaller birds. Well camouflaged, it is able to conceal itself in dense forest. The species' main call is a loud, piercing, rising screech or shriek, *zzhweeeesht*, often followed by a strident *pe-pe-pe-pe-pe-pe-pe*.

Habitat: Native or exotic forest or scrub holding one or more of their three host species. Present in Aotearoa only in spring, summer and autumn, on Te Hauturu-o-Toi, and from Waikato south to Ata Whenua Fiordland and Rakiura Stewart Island. During winter, they occur in an arc extending 11,000km from Palau in western Micronesia to Henderson Island in eastern Polynesia.

Ornithologist's notes: Like the pīpīwharauroa (page 33), koekoeā are brood parasites. Females lay their brownish-speckled white eggs mostly in November and December, in the nests of mohoua (page 121) or pīpipi (brown creepers, *Mohoua novaeseelandiae*) in Te Waipounamu South Island, and pōpokatea (page 119) in Te Ika-a-Māui North Island. After this, the adult cuckoos take no further part in breeding. Young cuckoos are dependent on their foster parents for several weeks after fledging, but many details of the eggs and young are poorly known. The birds eat cicadas, wētā, stick insects, skinks, and eggs and nestlings from other birds' nests.

Status in Aotearoa: Endemic

Conservation status: Nationally vulnerable

WEKA

Gallirallus australis

This large, furtive, flightless rail, known to early Pākehā settlers as the woodhen, has become extinct over much of Aotearoa but still thrives on Kawau, Mokoia and Kapiti Islands, Te Wharawhara Ulva Island, Rēkohu Chatham Islands, and various localities in upper Te Ika-a-Māui North Island and north-west Te Waipounamu South Island. Weka have red eyes, strong red legs, and a strong, pointed reddish to greyish bill. There are four subspecies: the North Island weka, which is redominantly grey-breasted, with a grey bill and brown legs; and the western, Stewart Island and buff weka (the last is extinct in eastern South Island but introduced and thriving on Rēkohu, where it is hunted for food), which have a grey to brown-grey breast with a wide brown breast-band, a grey to pink bill, and brown to pink legs. All birds measure 50–60cm in length and vary in weight from 430g to 1400g. Pairs make spacing calls at dawn and just after sunset: a characteristic *coo ... eet*. The male's call is lower and slower than the female's. Birds also boom and cluck.

Habitat: A wide variety of habitats, from the coastline to above the treeline, including wetlands, rough pasture, shrubland, and native and plantation forests.

Ornithologist's notes: One of eight native rail species in Aotearoa, this famously inquisitive snatcher of objects is an important disperser of forest seeds. Monogamous, some pairs produce multiple clutches each year. The nest is a woven cup of fine grass or sedge in dense vegetation, usually under an object or in a burrow. Females lay 1–4 large eggs, and incubate them mainly from early morning to late afternoon. The male does the rest, and most of the parental care. Omnivorous, weka eat forest fruits, invertebrates, eggs, skinks, small mammals, birds and carrion. They may also kill stoats and large rats.

Status in Aotearoa: Endemic

Conservation status: Not threatened

TAKAHĒ
SOUTH ISLAND TAKAHĒ

Porphyrio hochstetteri

A rare relict of the flightless birds that once ranged across Aotearoa, the takahē was considered extinct until famously being rediscovered at Te Puhi-a-noa Murchison Mountains of Ata Whenua Fiordland in 1948. In the 1980s takahē were translocated from there to islands and mainland sites around the country. The taller moho (North Island takahē, *Porphyrio mantelli*) was hunted by Māori; last seen in 1894, it is now extinct. The largest living rail species in the world (50cm long; 2–3.5kg), the takahē has vivid cobalt to turquoise blues on the head, neck and underparts, mossy greens on the wings and back, and a white undertail. The huge conical bill is rātā-flower red, and extends onto the forehead as a red frontal shield. The stout legs are red and orange. The bird's main calls are a loud shriek, a quiet hooting contact call and a muted boom of alarm. An alternative name is *Notornis*, the old Latin genus name.

Habitat: In the wild, alpine tussock grasslands with some time spent in beech forest. There is a captive takahē breeding facility in Te Anau, and free-ranging populations on nature reserves on both main islands and some offshore islands. As of 2022 there are 400 birds.

Ornithologist's notes: Takahē live in pairs or small family groups. Monogamous pairs defend a breeding territory by calling or fighting, returning to the same areas each year. The sooty young stay with their parents until just before the next breeding season, or for a second year. The females lay 1–3 large blotched eggs from late October to January; both sexes share incubation and chick-rearing. The chicks stay in the nest for a week; as they get stronger, they follow their parents, begging for food. Takahē feed on tussock grasses, sedges and rushes, and also eat pasture grasses, large insects and, rarely, ducklings and skinks.

Status in Aotearoa: Endemic

Conservation status: Nationally vulnerable

PŪKEKO

Porphyrio melanotus melanotus

This abundant, large (38–50cm long; 900–1090g) rail species is found throughout Aotearoa and has a reputation for being cheeky. The species can be hunted under licence and the blue breast feathers are prized for use in kākahu, or cloaks. The head, breast and throat are deep blue-violet, the back and wings are black, and the undertail is white. The conical red bill extends into a frontal shield on the forehead. The eyes are red and the legs and feet are orange, with long, slim toes. Similarly coloured, females are smaller than males and juveniles are duller. The main call is a loud territorial crowing. The various contact calls include *n'yip*, *hiccup* and *squawk*, and the defence call is a loud, shrill screech.

Habitat: Near sheltered freshwater or brackish swamps, streams and lagoons around Aotearoa, especially by open grassy areas, roadsides and drainage ditches, and margins of scrub or forested areas, from sea level to 2300m.

Ornithologist's notes: Pūkeko live in permanent social groups that defend a shared territory. Birds nest as monogamous pairs, or as a pair with an extra male or, more rarely, an extra female. These groups may also have non-breeding helpers. When multiple breeding females are present, they lay in the same nest. The clutch usually comprises 4–6 large eggs per female, and when all the females in a group use a single nest they can lay up to eighteen eggs. Breeding males incubate the eggs with some help from breeding females. All group members care for the chicks. After leaving the nest, the chicks are fed by the adults for two months. Laying peaks from August to January. Pūkeko are mainly herbivorous, but insects and worms make up a small part of their diet and, rarely, frogs, skinks, fish and nestling birds.

Status in Aotearoa: Native

Conservation status: Not threatened

TŌREA PANGO
VARIABLE OYSTERCATCHER

Haematopus unicolor

This large (48cm long; 720g), heavily built wading bird is long-lived, some reaching 30-plus years of age. The different colour morphs (black, intermediate or 'smudgy', and black and white or pied) were once thought to be different species or hybrids. This confusion was compounded by the way the proportion of all-black birds increases from north to south. These colour morphs interbreed and are now recognised as being the same single species – hence the inclusion of 'variable' in the common English name. All adults have a long orange bill, chunky coral-red legs, and red eyes with an orange eye-ring. Tōrea pango are very vocal, piping loudly in territorial interactions and when alarmed. The loud *ki-woo* flight call is similar to that of other oystercatcher species. Chicks are warned of danger with a sharp, loud *chip* or *click*.

Habitat: On beaches and in estuaries, and sometimes a short distance inland. Birds breed in monogamous pairs on sandy beaches, sandspits, dunes, shell banks, rocky shorelines and gravel beaches.

Ornithologist's notes: Along with the tōrea (South Island pied oystercatcher, *Haematopus finschi*) and tōrea tai (Chatham Island oystercatcher, *H. chathamensis*), this is one of 14 native waders that breed in Aotearoa. The nest is normally a simple scrape in the sand, and the 2–3 eggs are usually laid from October. The chicks are defended by both parents, but unlike most wading birds, tōrea pango feed their young. The growing chicks often beg from their parents well after fledging. The birds eat a wide range of coastal invertebrates, favouring shellfish such as mussels, tuatua and tuangi (cockles), which they open by pushing the tip of their bill between the shell and twisting, or by hammering. They occasionally eat oysters and small fish, and on grass they consume a range of invertebrates, including earthworms.

Status in Aotearoa: Endemic

Conservation status: Recovering

POAKA
PIED STILT

Himantopus himantopus

This elegant, medium to large (35cm long; 190g) wading bird is common in wetlands and coastal areas throughout Aotearoa. The plumage is white with black on the back of the neck and the upper wings, the long 'stilt' legs are pink, the eyes are dark, and the long, fine bill is black. Poaka tend to steer clear of people, making a shrill 'yapping' alarm call when approached. A less strident version of this call is made by flocks flying at night.

Habitat: All types of wetlands, from brackish estuaries and saltmarshes to freshwater lakes, swamps and braided rivers, where they feed in shallow water or mud and roost in shallow water or on banks. After breeding, they migrate towards more northerly coastal locations.

Ornithologist's notes: One of two stilt species that breed in Aotearoa, the other being the kakī (black stilt, *Himantopus novaezelandiae*). Poaka and kakī hybridise, which threatens the survival of kakī as a species because only about 170 birds remain. Kakī was not recognised as a separate species until the 1920s. Birds form groups throughout the year, feeding and roosting together in what can become large, noisy flocks, usually alongside other waders such as oystercatchers and godwits. Monogamous pairs breed during July–October, usually in colonies of about fifteen pairs, although some comprise up to 100 pairs. Both parents build the nest, which measures up to a few centimetres high, using mud, vegetation and debris, and siting it on the ground near water. They both incubate the 3–4 blotchy, light brown eggs for 25 days. The chicks can fly about a month after hatching, but the parents continue to care for them until the end of the season. The diet comprises insects, larvae and worms.

Status in Aotearoa: Native

Conservation status: Not threatened

TŪTURIWHATU
NEW ZEALAND DOTTEREL

Charadrius obscurus aquilonius

A bold, stocky wading bird (25cm long; 146–160g), tūturiwhatu is seen on sandy east coast beaches in the northern North Island and is sparsely distributed elsewhere. Of the two widely separated subspecies, the northern is more numerous. The larger, heavier and darker southern subspecies was once widespread in the South Island, but now breeds only in subalpine herbfields on Rakiura Stewart Island and winters on the Rakiura and Murihiku Southland coasts. In breeding plumage, the breast is orange-red in northern birds and brick red in southern birds. The head and back are brownish, and the underparts are off-white, with males generally darker than females. The robust bill is black and the legs mid-grey. The main call is a sharp *chip*, which is used as a contact call and increases in frequency as the perceived threat level rises. A high-pitched *tseep* warns chicks to hide, and a long, rattling *churr* is used when chasing intruders.

Habitat: Northern birds breed on sandy or gravel beaches, sandspits and shell banks. Southern birds breed in subalpine herbfields or rocky areas above the treeline on Rakiura, and after the breeding season fly to the Murihiku coast, where they spend the winter in a single flock.

Ornithologist's notes: The southern subspecies *obscurus* has a population of only about 200 and is considered by some authorities to be a separate species. Tūturiwhatu breed in monogamous pairs that defend territories. Northern birds nest in scrapes of sand, gravel or shells, sometimes sparsely lined. Females lay three eggs from August. Southern males do most of the incubation night shifts. Following the month-long incubation, the chicks fledge after six weeks. Southern birds nest in hollows in cushion plants or between rocks lined with tussock grass. Tūturiwhatu eat small insects, worms, fish, mussels and crabs. The species has more than twenty Māori names.

Status in Aotearoa: Endemic

Conservation status: Recovering

NGUTU PARE
WRYBILL

Anarhynchus frontalis

This very distinctive small (20cm long; 55g), pale plover is the only bird species in the world with a bill that curves to the right, which it uses to catch insect larvae from under riverbed stones and to sift through silt for tiny crustaceans. Mainly pale grey in all plumages, ngutu pare is perfectly camouflaged among the greywacke shingle of braided riverbeds. Its forehead, breast and belly are white, with a black upper breast-band from midwinter to the end of the breeding season that is more prominent in males. The long, pointed black bill curves to the right and the legs are grey-black. The sexes are alike in non-breeding plumage; juveniles lack a breast-band. The main call is a *chip*, which signals alertness. A rapid *churr* is made when chasing intruders, and a very quiet grating call is used to call to chicks.

Habitat: Breeds on Te Waipounamu South Island braided riverbeds and winters on inter-tidal mudflats in harbours and estuaries. High-water roosts are usually near foraging areas on local shell banks and beaches, and sometimes on pasture. Migrating flocks mainly follow coastlines, with most birds passing through Te Waihora Lake Ellesmere to and from breeding grounds.

Ornithologist's notes: After breeding, almost all birds migrate north to winter in northern harbours, mainly Tīkapa Moana o Hauraki Firth of Thames and Te Mānukanuka-o-Hoturoa Manukau Harbour. At this time, large flocks coordinate aerial manoeuvres or murmurations. Monogamous dispersed pairs defend territories. The male uses its breast to bulldoze a shallow scrape in gravel, then lines it with small stones. Clutches of two eggs are laid from September. Incubation is shared for 30–36 days. The chicks are guarded by one or both parents during their first three weeks, before fledging after 35–40 days. The birds eat mayfly and caddis-fly larvae, insects, worms, and small crustaceans and fish.

Status in Aotearoa: Endemic

Conservation status: Nationally increasing

SPUR-WINGED PLOVER

Vanellus miles novaehollandiae

This tall, stocky wader (38cm long; 370g) is a common self-introduced native species, originally from Australia. It was first recorded breeding in Aotearoa near Waihōpai Invercargill in 1932 and has since spread northwards through the country, reaching Te Tai Tokerau Northland in the 1980s. It is a conspicuous, noisy bird of open country. The spur-winged plover has a yellow bill, wattles and eye-rings; a black crown and collar; a white head, neck and front; and a grey-brown back with a white rump. The black tail has a thin white tip and the long legs are reddish brown. It is named after the thorny yellow spurs on its wings, which it displays to threaten predators but rarely uses. It makes a loud, shrill, staccato rattle, often at night.

Habitat: Low vegetation, often near water, from the margins of marine and terrestrial wetlands, riverbeds and lake shores, to estuaries, beaches, farm pastures, and grassland in urban areas, parks and sports fields.

Ornithologist's notes: There are two subspecies in Australia. The southern subspecies, which is the one found in Aotearoa, is recognised by some authorities as a separate species, the black-shouldered lapwing (*Vanellus novaehollandiae*). Adults breed in monogamous isolated pairs that share incubation and chick care. The nest is in a wide range of open habitats, but most commonly in areas associated with human activities, including pasture, cropland, urban parks and golf courses. Most nests are a simple scrape lined with dried grass, twigs or small pebbles. The female lays 3–4 olive-brown eggs with dark spots and blotches, usually in June–November. Incubation takes 30–34 days, with chicks leaving the nest soon after hatching, fledging at 6–7 weeks and independent after 8–9 months. Birds eat a wide range of marine and terrestrial invertebrates, including molluscs, crustaceans, insects and worms.

Status in Aotearoa: Native

Conservation status: Not threatened

KUAKA
BAR-TAILED GODWIT

Limosa lapponica

This imposing, large (39–41cm long; 275–600g), long-legged wader can be seen on many estuaries and harbours in summer, with the main populations at Pārengarenga Harbour, Kaipara Harbour, Te Mānukanuka-o-Hoturoa Manukau Harbour, Tīkapa Moana o Hauraki Firth of Thames and Onetāhua Farewell Spit. It is brown above and pale below, and has a long, tapering, slightly upturned pink bill with a black tip. Males are markedly smaller, with shorter bills than females. Kuaka make the longest non-stop migration flight (8–9 days, from Aotearoa to western Alaska) of any non-seabird, but unlike seabirds, they do not stop to feed. During its average 15-year lifespan, an individual kuaka flies at least 385,000km, the distance from the Earth to the Moon. Kuaka usually call in flight, *a-wik, a-wik, a-wik* or *kua-ka* (hence their Māori name, which is also used for the common diving petrel, *Pelecanoides urinatrix*), but for most of their time in Aotearoa are silent on the ground. This changes just before migration in March each year, when there is a marked increase in the frequency and volume of calls from birds about to leave.

Habitat: Sandy beaches, tidal mudflats and wetlands, marshes and sometimes on wet pastures.

Ornithologist's notes: One of about a dozen Arctic-breeding wader species that migrate annually to Aotearoa. Females lay four large eggs in a shallow bowl lined with lichen at their Arctic breeding grounds. Monogamous pairs share incubation and hatching duties, although one parent may depart earlier for their migration staging area. Birds do not breed until their third or fourth year, so each southern winter hundreds of non-breeding individuals remain in Aotearoa. Kuaka eat worms, shellfish and crustaceans in Aotearoa, and at their Arctic breeding grounds they feed on cranefly larvae, other invertebrates and berries.

Status in Aotearoa: Native

Conservation status: Declining

TUTUKIWI
SUBANTARCTIC SNIPE

Coenocorypha aucklandica

This well-camouflaged small wader (23cm long; 110g) lives among dense vegetation on the subantarctic Motu Maha Auckland Islands, Moutere Mahue Antipodes Islands and Motu Ihupuku Campbell Island, each of which supports a separate subspecies. The Motu Ihupuku subspecies was discovered on a small island in the group in 1997. Following the eradication of rats on the main island in 2001, this nationally vulnerable subspecies naturally recolonised it, completing a remarkable story of discovery and recovery within less than a decade. Adults have cryptic brown plumage, stout legs and a slender bill that is about 5cm long. The head has black and reddish-brown stripes and the body is brown, mottled with black and reddish brown. The main call is a strident *chup chup*, mostly given at night. The species has a nocturnal courtship display, called hakawai, in which males make vertical dives at speed from considerable heights, accompanied by vocal sounds and a roar made by the vibrating tail feathers.

Habitat: Tussock grasslands, herbfields, ferns, shrubland and low native forest.

Ornithologist's notes: New Zealand's *Coenocorypha* snipes are possibly a relict taxon of an ancient lineage, older than the larger *Gallinago* snipes of Australia, Asia, Africa, Europe and the Americas. Tini Heke Snares Islands and Rēkohu Chatham Islands also have separate endemic snipe species. The North Island snipe (*C. barrierensis*) and South Island snipe (*C. iredalei*) became extinct in 1870 and 1964, respectively. The breeding habits of tutukiwi are poorly known. All nests found contained one or two olive-brown eggs with darker spots and blotches. Monogamous pairs share incubation, and it is thought that each adult cares for one chick independent of its mate. The birds feed on soil-dwelling invertebrates by probing with their long, narrow bills.

Status in Aotearoa: Endemic

Conservation status: Naturally uncommon

TARĀPUNGA
RED-BILLED GULL

Chroicocephalus novaehollandiae scopulinus

The most familiar and abundant gull species of the Aotearoa coast, tarāpunga gather in large, noisy flocks year-round to breed, feed and roost. Adults are almost completely white, but the back and upper wings are pale silver-grey, and the main flight feathers are black with white tips. The iris is white, and the bill, eyelids and feet are bright red in the breeding season but duller during the rest of the year. Both sexes are similar in size (37cm long; 240–320g), but males are slightly larger and have a longer, heavier bill. Immatures are similar to adults, but with brown patches on the back and brownish flight feathers. Birds use a wide range of calls, most notably a strident *kek* during breeding season.

Habitat: Coastal, foraging up to 1km offshore. Rarely found inland, except at Te Rotoruanui-a-Kahumatamomoe Lake Rotorua, and more often seen in coastal towns at beaches, landfills, and fishworks or meatworks. Breeds in dense colonies, mostly on eastern coasts at rock stacks, cliffs, river mouths, and sandy and rocky shores. The largest colonies – at Kaikōura, Manawatāwhi Three Kings Islands and the Mokohinau Islands – have declined markedly over recent decades.

Ornithologist's notes: One of three gull species in Aotearoa, tarāpunga breed in large, dense colonies on the mainland, and in scattered, concealed sites on subantarctic islands to avoid egg and chick predation by brown skuas (*Catharacta antarctica*). Egg-laying is from September to January. Birds are monogamous, and both sexes share in nest-building, incubation and chick-feeding. The nest is made from grass, seaweed or twigs. The clutch usually comprises two brown or green-grey eggs with dark spots and blotches. Incubation lasts about three weeks and chicks begin to fly after a month, and are then fed by their parents for about another month. Tarāpunga eat mainly krill, worms, small fish, offal and flies.

Status in Aotearoa: Native

Conservation status: Declining

TARA ITI
NEW ZEALAND FAIRY TERN

Sternula nereis davisae

This small, delicate coastal tern (25cm long; 70g) is our most endangered endemic bird. The relict population of fewer than forty birds, including ten breeding pairs, survives between Whangārei and Tāmaki Makaurau Auckland, with eight chicks fledging in the 2021–2022 breeding season. Threatened by introduced predators and disturbance by humans, tara iti are intensively managed during the breeding season, including the captive rearing of some chicks for release into the wild. A DNA study comparing this subspecies with larger breeding populations of different subspecies in Australia (*Sternula nereis nereis*) and New Caledonia (*S. n. exsul*) found that they are genetically distinct: the estimated gene flow was low to zero, indicating no interbreeding between them. Breeding adults are white with pearly-grey upperparts, a yellow-orange bill and orange legs. A black cap covers the crown and nape, extending forward to surround each eye and forming an irregular patch in front that does not quite reach the bill. A rounded white 'notch' projects into the black cap above the eyes and connects with the white forehead. Birds make repeated high-pitched calls, including *tiet, tiet* and *kek, kek,* and have an alarm call, *zipt-zipt-zipt*.

Habitat: Breeds successfully at only a few sandspits on the Te Tai Tokerau Northland coast, with birds feeding in adjacent estuaries and offshore. Winters in Kaipara Harbour.

Ornithologist's notes: One of eight tern species that breed in Aotearoa. Monogamous pairs breed in October–March. Females lay 1–2 speckled creamy eggs in an unlined scrape nest and do most of the incubation. Both parents guard and feed the chicks, with the male providing most of the food. Chicks fledge after a month and are fed by the parents with reducing frequency for another month. The birds eat small fish such as juvenile flounder, gobies, elvers and shrimps.

Status in Aotearoa: Native

Conservation status: Nationally critical

TARAPIROHE
BLACK-FRONTED TERN

Chlidonias albostriatus

This dainty, medium to small (28cm long; 95g) marsh tern is often seen in distinctive buoyant flight over braided rivers or pasture, or sometimes subalpine tussocks, catching insects in the air or skinks and worms off the ground. Early Pākehā colonists nicknamed tarapirohe 'plough boys' or 'plough birds' after their habit of following ploughs to glean worms. Adults have long, pointed wings and a modestly forked tail. In breeding birds, the slate-grey plumage contrasts with the black cap, white 'cheeks', short orange legs and decurved, bright orange bill. Non-breeding adults have a mottled grey cap, a black patch around the eye and 'ears', and a black-tipped bill. Juvenile and immature birds are similar to non-breeding adults but with a darker bill. Noisy at colonies, birds call harshly, especially when diving at intruders, which they sometimes strike. Their usual call is a repetitive *kit*.

Habitat: Te Waipounamu South Island braided rivers and adjacent pasture or tussock, coastal estuaries and lagoons, coastal farmland and over waters up to 50km offshore. Breeds on braided riverbeds and estuaries in Te Waipounamu, from sea level up into the high country. After breeding, birds disperse to the coast, with some migrating to Te Ika-a-Māui North Island wintering sites in Wairarapa, Te Matau-a-Māui Hawke's Bay, Te Moana-a-Toi Bay of Plenty and Kaipara Harbour.

Ornithologist's notes: Tarapirohe are monogamous and usually breed in colonies of fewer than fifty pairs, but some have up to 250. Nests are a simple scrape in sand or gravel lined with twigs. Females lay 1–2 eggs from October to December. Incubation is shared and lasts twenty-five days. Chicks leave the nest a few days after hatching, with the parents feeding them invertebrates, small fish and skinks. They fledge after a month and continue to be fed for a further few weeks. Adults eat stonefly larvae, small fish, worms and skinks.

Status in Aotearoa: Endemic

Conservation status: Nationally endangered

TARA
WHITE-FRONTED TERN

Sterna striata

This graceful, medium-sized tern (42cm long; 160g) is the commonest on New Zealand's coasts, at times flocking in the hundreds or thousands. The English name 'white-fronted' refers to the forehead, where a thin strip of white feathers separates the black cap from the black bill. Adults are very pale grey above and white below, with pointed wings and a long forked tail. The black cap of breeding adults extends down to the back of the neck; non-breeding adults have a reduced cap. The sexes are similar but females are slightly smaller. Immature birds are similar to non-breeding adults but with a white-streaked black nape and upper wings, and light brown mottling. Newly fledged birds have fine blackish barring over the back and wings. The main call is a high-pitched *siet*, and a harsh *keark* is made when defending nests.

Habitat: Coasts of the main islands and outlying islands, and on larger rivers in Waitaha Canterbury. Mainly marine, but seldom far from the coast. In autumn, some birds (mostly immatures) migrate to Australia.

Ornithologist's notes: One of eight tern species that breed in Aotearoa. Monogamous pairs usually breed in large colonies on shingle riverbeds, sand dunes, rock stacks and cliffs. Egg-laying is highly synchronised within a colony. Females usually lay a single speckled creamy egg, in a slight depression on the ground with no nesting material except a few small stones to line the nest. Incubation is shared, with the egg hatching after 24 days and the chick fledging after fifty days. The birds feed on small fish and elvers taken at sea, in lagoons or in rivers. At sea they often feed with flocks of gulls and shearwaters over shoaling fish, including kahawai, hence the species' alternative common name of kahawai bird. Tara is also used as a general name for terns.

Status in Aotearoa: Native

Conservation status: Declining

AMOKURA
RED-TAILED TROPICBIRD

Phaethon rubricauda

Amokura are elegant, large (46cm long; 800g) white tern-like tropical seabirds with two long, thin red tail streamers (up to 40cm), small black eye patches, black webbed feet and a heavy, pointed red bill. Truly oceanic, they remain at sea almost constantly, returning to land only to breed. They fly with slow, regular wingbeats, alternating with gliding, and have been observed briefly flying backwards. The two red streamers grow from the centre of the wedge-shaped tail but wear out quickly, and so are constantly replaced. Immature birds are barred black and white, with black on the outer wing feathers and a dark grey-reddish bill, and lack tail streamers. Tropicbirds make harsh, guttural calls, trumpet-like growls and snarls, and loud cackling.

Habitat: Breeds in Aotearoa only in subtropical Rangitāhua Kermadec Islands, on Raoul, North Meyer, South Meyer, Nugent, Macauley, Curtis, Dayrell and South Chanter Islands. Feeds over the open ocean, seeking cool, low-salinity, nutrient-rich waters. Nest sites are typically in coastal cliff crevices or on ledges that have some sheltering vegetation.

Ornithologist's notes: Monogamous pairs breed at a solitary nest or in loose colonies if a suitable site is available. Breeding takes place from December to August. At Rangitāhua, egg-laying typically peaks in December and January, with most nests containing young from mid-January to May. Birds dig the nest scrape with their feet and defend it as a territory that the pair reuses across years. Females lay a single pale egg with dark purple-brown flecks. Both adults share incubation and chick-feeding. Amokura eat mainly small to medium-sized fish, flying fish, squid and flying squid, which they catch by plunge-diving.

Status in Aotearoa: Native

Conservation status: Recovering

HOIHO
YELLOW-EYED PENGUIN

Megadyptes antipodes

The largest penguin species that breeds on the mainland and an icon of conservation, the hoiho is also the only penguin with yellow eyes. It is the world's second-rarest penguin species (after the Galápagos penguin, *Spheniscus mendiculus*). The hoiho is a tall, stocky penguin (65cm long; 5kg) with a pale yellow band of feathers extending from each yellow eye around the back of the head, a robust red-brown bill and mainly pink feet. The rest of the head, neck, back and upper side of the flippers are slate blue, while the breast, belly and underside of the flippers are white. Both sexes look alike, but males are larger. Juveniles lack the yellow band and have paler eyes and a paler head. At the nest site, birds make a loud, high-pitched 'braying' (hoiho means 'noise shouter'). The species is the only member of the genus *Megadyptes*, which means 'large diver'.

Habitat: Breeds in coastal native forest, regenerating coastal scrub, tussocks, pasture, under planted exotic trees and on cliffs.

Ornithologist's notes: One of six penguin species that breed in Aotearoa, the hoiho selects well-concealed nest sites under dense vegetation from July. Both sexes build the nest, which is a shallow bowl of twigs, grass and leaves. The mean lay date is 24 September on Muaupoko Otago Peninsula, but later further south. Females lay two large white eggs. Both sexes incubate the eggs for 39–51 days, with most hatching in November. Both parents guard and feed the chicks for six weeks, and continue to feed them during the post-guard phase until about 106 days from hatching, when the chicks are left alone at the nest during the day. Fledging takes place from February to as late as April. Birds forage up to 25km offshore, diving to depths of up to 40–120m for bottom-dwelling fish such as rawaru (blue cod), kupae (sprats), hoka (red cod) and kohikohi (opalfish), squid and crustaceans.

Status in Aotearoa: Endemic

Conservation status: Nationally endangered

KORORĀ
NEW ZEALAND LITTLE PENGUIN

Eudyptula minor minor

As its English name suggests, this is the world's smallest penguin species, at 33cm and weighing just over 1kg. The most common penguin on the mainland, it breeds from Te Tai Tokerau Northland to Rakiura Stewart Island and Rēkohu Chatham Islands, and around the mainland coast. Kororā are deep blue to slate blue with a white throat, breast and belly. They have a straight dark bill with a hooked tip, blue-grey or hazel eyes, and pink legs and feet. Males are slightly larger than females. Birds on Te Pātaka-o-Rākaihautū Banks Peninsula have distinctive white-bordered flippers. When coming ashore at night, kororā make a range of growls, screams, cat-like mews and trumpeting, and a contact 'bark' at sea.

Habitat: Common along most coastlines – especially on offshore islands, which offer greater protection. The main breeding areas include Tīkapa Moana Hauraki Gulf, Te Whanganui-a-Tara Wellington, Te Tauihu-o-te-waka Marlborough Sounds, Te Pātaka-o-Rākaihautū, Oamaru and Muaupoko Otago Peninsula.

Ornithologist's notes: Most closely related to the Australian fairy penguin, another subspecies of little penguin (*Eudyptula minor novaehollandiae*). Birds nest close to the sea in burrows, caves and rock crevices, or under logs or built structures such as nest boxes, pipes, wood piles and baches. The nest is often lined with sticks and seaweed. They can breed as isolated pairs, in colonies or semi-colonially. Monogamous within a breeding season, both adults share incubation and chick-rearing. Females lay 1–2 white to lightly mottled brown eggs between July and November. Incubation takes up to thirty-six days. The chicks are fed by the parents for about a month, and fledge after about two months. During breeding adults forage within 20km of the colony, diving down to 35m to catch small fish and squid.

Status in Aotearoa: Native

Conservation status: Declining

TOROA
NORTHERN ROYAL ALBATROSS

Diomedea sanfordi

This is one of two royal albatross species that breed only in Aotearoa, the other being the southern royal albatross (*Diomedea epomophora*). Along with the wandering albatross (*Dimoedea exulans*, also called toroa), it is the largest seabird in the world, with a wingspan of 3m or more and weighing 6–9kg (males are larger than females). Adults have a white body and tail; long, narrow black upperwings; white underwings; a huge, pale pink hooked bill; and pale pink legs and webbed feet. Birds are long-lived, reaching 30 years on average, with one recorded at over 60 years. Pairs breed on remote islands in Rēkohu Chatham Islands and at Pukekura Taiaroa Head near Ōtepoti Dunedin, which is accessible to the public. Toroa is also used as a general name for albatrosses.

Habitat: Spends most of its time at sea, mainly as a solitary forager over continental shelves and shelf edges. During the breeding season mostly forages over the Chatham Rise. Non-breeding and immature birds embark on a downwind circumnavigation in Te Moana Tāpokopoko a Tāwhaki Southern Ocean.

Ornithologist's notes: One of twelve albatross species that breed in Aotearoa – more than in any other country. It has an elaborate courtship display, involving bowing, bill-fencing, dancing with outstretched wings, 'skycalling' (both birds hold their head back and point their bill at the sky) and loud braying. Pairs form long-term monogamous bonds, usually for life, and share incubation and chick-rearing. They nest in colonies and breed in alternate years. A breeding cycle takes almost a whole year to complete, with the single huge white egg laid in November. Hatching takes place in January or February after about two months of incubation. The chick is guarded and fed, before fledging at about eight months in September. The birds eat mainly squid, fish and crustaceans.

Status in Aotearoa: Endemic

Conservation status: Nationally vulnerable

TĪTĪ
SOOTY SHEARWATER

Ardenna grisea

This large (45cm long; 650–950g), dark seabird is common in coastal Aotearoa south of Ōtautahi Christchurch, often gathering at sea in spectacular flocks of tens of thousands to feed on bait balls of crustaceans. It frequently dives to depths of 16m, but sometimes to more than 60m. It has long, narrow wings; a long, slender hooked bill; and a narrow, short tail. The upperparts are sooty brown, the underparts are slightly greyer with a silver-grey flash on the underwings, and the feet are dark grey. Most calls are made at night by birds in breeding colonies. The main call is a loud, rhythmically repeated, slightly frenzied *coo-roo-ah*. Tītī is also used as a general name for petrel species.

Habitat: Present on many islands, from Manawatāwhi Three Kings Islands in the north to the subantarctic islands, but all large colonies are on Rakiura Stewart Island or Tini Heke Snares Island.

Ornithologist's notes: One of seven shearwater species that breed in Aotearoa. Birds breed annually from September to May, often with the same mate, in large, dense colonies. Afterwards, they migrate to the North Pacific Ocean, following a clockwise circular route, stopping over in Japan, Alaska or California before returning for the next breeding season. A flock of 'crazed' sooty shearwaters seen on the California coast in 1961, reportedly flying into objects and dying on the streets, inspired Alfred Hitchcock's film *The Bird*s (1963). It is likely that toxin-forming algae were present in plankton the shearwaters had eaten. Monogamous pairs share incubation and chick care. Females lay a single large white egg in November or December in a nest chamber at the end of a burrow. Also known as muttonbirds, the young are harvested from burrows at traditional sites around Rakiura. Tītī eat fish, squid, and krill, and offal from fishing vessels.

Status in Aotearoa: Native

Conservation status: Declining

PAKAHĀ
FLUTTERING SHEARWATER

Puffinus gavia

Named for its distinctive fluttering flight, the pakahā is a common seabird of inshore waters in the north-east of Te Ika-a-Māui North Island, Te Tauihu Marlborough and Te Whanganui-a-Tara Wellington. Fast-moving foraging flocks, sometimes comprising thousands of birds, are a spectacular sight as they pursue schools of small fish. During this frenetic activity, many of the birds splash into the water and dive below the surface, resurfacing and repeating the action. Pakahā are small (37cm long; 365g), with a dark cap, dark brown upperparts, white underparts and dusky underwings with variable dark markings. The long, thin, dark bill is hooked, and the legs and feet are pinkish brown. The birds make a distinctive staccato call, *ka-hek-ka-hek-ka-he*k, mostly in flight over colonies.

Habitat: Offshore islands throughout northern Aotearoa and some islands in Te Tauihu-o-te-waka Marlborough Sounds while breeding. After breeding, large numbers remain within local inshore waters through the winter months. Between February and August, migrants cross Te Tai-o-Rehua Tasman Sea to eastern and south-eastern Australian waters.

Ornithologist's notes: This species is surprisingly poorly known for one that is so abundant. Monogamous pairs nest in short burrows under scrub or forest, with females laying a single large white egg in September–October. Incubation and chick care are shared, but incubation and nestling periods are not fully known. Chicks fledge in January in the north and in February further south. Pakahā eat small fish such as mohimohi (pilchards) and kupae (sprats), and krill. They catch fish by pursuit diving, using partially folded wings to swim underwater, and often catch krill by 'snorkelling' forward at the surface with wings raised and head submerged to search underwater.

Status in Aotearoa: Endemic

Conservation status: Relict

TĀKAPU
AUSTRALASIAN GANNET

Morus serrator

Tākapu are temperate-water relatives of the smaller tropical boobies. Often seen flying over Aotearoa coastal waters in summer, they are best known for their impressive 1.8m wingspan and dramatic torpedo-like plunge dives, made from up to 15m above the sea. These seabirds are large and heavy (90cm long; 2.3kg) yet streamlined, with mainly white plumage, a buff-yellow head and neck, bright blue skin around the eyes and a long, tapering bluish-grey bill. Adult males and females are similar, juveniles have mottled brown-and-white plumage, and immatures gradually acquire whiter plumage until they reach maturity at three years of age. The birds announce their landing with a distinctive *urrah* call, while a shorter *oo-ah* call indicates take-off.

Habitat: Dense, teeming colonies on coastal islands, and on cliffs and beaches of some headlands on the mainland. The largest mainland gannetry is at Te Kauwae-a-Māui Cape Kidnappers, where 5000 pairs breed. Other mainland breeding sites include Muriwai and Onetāhua Farewell Spit. The juveniles disperse swiftly to eastern and southern Australian waters, returning to Aotearoa to breed 3–7 years later.

Ornithologist's notes: These birds can plunge-dive 1–2m underwater, then dive deeper propelled by their wings and large webbed feet. The impact of the dive is cushioned by inflatable air sacs in the neck and breast, and a reinforced skull. Birds are rather ungainly on the ground. The breeding season extends from July, when birds first return to their gannetries, to chick fledging in March–April. Monogamous breeding pairs form and maintain a nest mound. Courtship displays include both birds bowing and pointing their heads skywards, and mutual preening. Females lay a single large, pale bluish-white egg. Both adults share incubation and then brood the chick on their webbed feet. Tākapu mainly eat pilchards, anchovies, barracouta, flying fish and squid.

Status in Aotearoa: Native

Conservation status: Not threatened

KĀRUHIRUHI
PIED SHAG

Phalacrocorax varius varius

A striking, large (65–85cm long; 1.3–2.1kg) white-fronted black shag with a vivid blue eye-ring, yellow facial skin, and pink-red skin at the base of the bill during the breeding season. Kāruhiruhi have a long neck; long, hooked, pale creamy-grey bill; green eyes; and black legs with webbed feet. The long neck is held forward during flight. They are often seen alone or in small groups roosting on rocky headlands, coastal trees or jetties. Usually silent away from their nesting colonies, they are vocal at colonies during pair formation and nest-building, and when they return to the nest during incubation. Females give wheezy *haa* calls and males give a variety of loud calls, sometimes repeated: *aark, kerlick* and *whee-eh-eh-eh*.

Habitat: Coastal marine waters, harbours and estuaries, and occasionally in freshwater lakes and ponds near the coast. Breeds on the coast in three main areas: in the north from Te Tai Tokerau Northland to Tāmaki Makaurau Auckland and west to Te Tairāwhiti East Cape; a central zone from Te Whanganui-a-Tara Wellington, Whakatū Nelson, Te Tauihu Marlborough and south to Waitaha Canterbury and Te Pātaka-o-Rākaihautū Banks Peninsula; and southern Te Waipounamu South Island and Rakiura Stewart Island.

Ornithologist's notes: One of twelve shag species that breed in Aotearoa – more than in any other country. Monogamous pairs mainly nest in trees near coastal cliffs, with a few colonies forming in trees by freshwater lakes near the coast. Females lay 2–5 large, pale blue-green eggs, usually in February–April and August–October, and both adults share incubation and chick care. The chicks start to fly at about a month old, staying near the colony to be fed by their parents for another 10–11 weeks. The birds eat fish and occasionally crustaceans.

Status in Aotearoa: Native

Conservation status: Recovering

KAWAU TIKITIKI
SPOTTED SHAG

Phalacrocorax punctatus

In breeding plumage, this is one of New Zealand's most colourful and striking native birds. A medium-sized shag (64–74cm long; 700–1200g), it looks flamboyant in its breeding finery, sporting a Mohican-like double crest with short white filoplumes growing from the neck and thighs, and a broad white stripe extending from above the eyes down both sides of the neck. It has distinctive pale blue-grey plumage with bright green-blue facial skin, a blue eye-ring, and yellow legs and webbed feet. The species name comes from the small black spots that appear near the tip of the back and wing feathers during the breeding season. The birds make loud grunts at their resting, roosting and nesting areas, but are otherwise silent.

Habitat: Te Waipounamu South Island, lower Te Ika-a-Māui North Island and Tīkapa Moana Hauraki Gulf coastal waters out to 16km, entering inlets and estuaries to feed and roost. Entirely marine, they breed on coasts of the North and South Islands and Rakiura Stewart Island. Outside the breeding season, they form large feeding and roosting flocks of up to 2000 birds.

Ornithologist's notes: Kawau tikitiki and the closely related Pitt Island shag (*Phalacrocorax featherstoni*) are the only yellow-footed shag species endemic to Aotearoa. They breed in scores of colonies, mostly in the South Island, of up to 700 pairs, with the timing of breeding varying according to location and food availability. Pairs are monogamous. Females lay 3–4 large, pale blue eggs in a nest platform made of sticks and vegetation and built 1m apart on coastal cliff ledges and stacks. The parents share incubation and chick-rearing, with juveniles leaving the nest after two months. The birds feed on small fish, squid and plankton.

Status in Aotearoa: Endemic

Conservation status: Nationally vulnerable

KŌTUKU
WHITE HERON

Ardea alba modesta

The kōtuku is a majestic 1m-tall white heron (700–1200g), with gleaming white plumage, a long yellow 'spear' bill, a long neck, yellow eyes and long dark legs. Breeding birds have a grey-black bill and develop up to fifty long filamentous plumes, mainly on the back, which they raise and fan out during breeding displays. In flight, the kōtuku tucks its head back, which obscures the length of its neck and gives it a hunched appearance. When walking, it has an elegant upright stance, showing the extreme length of its neck. The call is a harsh croak. The whakataukī 'He kōtuku rerenga tahi', 'A white heron's flight is seen but once', is used to refer to a very special or rare event.

Habitat: Harbours, estuaries, freshwater wetlands and high-country lakes, from Murihiku Southland to Te Tai Tokerau Northland outside the breeding season. The only breeding site in Aotearoa is in a kahikatea swamp near Ōkārito Lagoon, Te Tai Poutini Westland.

Ornithologist's notes: The largest of four heron species that breed in Aotearoa, kōtuku arrive at the breeding colony in August. Females lay 3–5 large white eggs between September and October in nests built in trees and tree ferns. Both parents incubate and feed the chicks. The breeding season at the lagoon coincides with the upstream migration of native īnanga (whitebait), which are the species' main food during nesting. Elsewhere, they eat small fish, frogs, skinks, invertebrates, small birds such as tauhou (page 133), and small rodents.

Status in Aotearoa: Native

Conservation status: Nationally critical

MATUKU MOANA
WHITE-FACED HERON

Egretta novaehollandiae novaehollandiae

This species is now the most common heron in Aotearoa, despite being a relatively recent self-introduced arrival from Australia in the 1940s. These elegant, medium-sized (67cm long; 550g) blue-grey herons can be seen stalking along the edge of almost any coastal or freshwater habitat. They have a white face and throat, a long dark 'spear' bill, and pale yellow legs. Their backs are blue-grey and the breast and underside are tinged maroon-brown. In breeding plumage they have long strap-like grey plumes on the back, and shorter pinkish-brown plumes on the breast, which they raise and fan out in courtship displays. They also perform aerial displays near the nest. In flight, the head is usually tucked back towards the shoulders and there is a marked contrast in the open wings between the dark grey main flight feathers and the rest of the pale grey wing feathers. The birds can also fly with their neck outstretched. Juvenile and immature birds are paler grey and lack the white throat. The herons make a harsh croak in flight.

Habitat: Mainly on rocky shores and estuary mudflats, but also found on lakesides up to 500m above sea level, in farm ponds, and on damp pastures and grassy sports fields.

Ornithologist's notes: Pairs are monogamous and usually nest in the tops of large pine or macrocarpa trees growing near water, or sometimes on artificial structures. Females lay 3–5 large white eggs in a large, loose platform nest, which both parents incubate for about twenty-six days. Chicks fledge at about forty-six days and it is unusual for more than two to be raised at a time. The diet comprises a wide range of prey, including small fish, crabs, worms, insects, spiders, tadpoles, frogs, skinks and mice. The Māori name for the species is also used for the reef heron (*Egretta sacra*).

Status in Aotearoa: Native

Conservation status: Not threatened

KĀHU
SWAMP HARRIER

Circus approximans

The largest of the world's 16 harrier species (50–60cm long; 650–850g), this impressive bird of prey has a wingspan of up to 1.5m and is an opportunistic hunter. Long-legged, with large-taloned feet, owl-like facial discs and a strongly hooked bill, the harrier's plumage is highly variable. Adult birds generally have a tawny-brown back; a brown-streaked, pale cream breast; yellow eyes; and a creamy-white rump. They become paler with successive moults, with the oldest birds having near-white undersides. Females are slightly larger than males, and juvenile and immature birds are uniformly dark chocolate brown. Kāhu most often search for food low to the ground in a gently rocking glide interspersed with shallow wingbeats. They glide or soar with their wings set in a shallow V-shape, with upturned fingered wingtips and an outspread tail. During courtship displays both birds perform spectacular rocking dives and then swoop back up in a large U-shaped loop, accompanied by a high-pitched *kee-o, kee-o* call.

Habitat: Coastal fringes, estuaries, wetlands, pine forest, farmland and high-country areas. Breeds in wetlands and areas of long grass and scrubby vegetation. Abundant throughout most of Aotearoa and on Rēkohu Chatham Islands, and often reaches distant offshore islands, from the subantarctic islands to Rangitāhua Kermadec Islands.

Ornithologist's notes: Kāhu are solitary monogamous breeders. The female lays 2–4 large white eggs between October and December, and incubates and feeds the chicks. The male passes food to the female in dramatic aerial handovers. A bulky stick nest is built on the ground or in low bushes, long grass, scrub or wetlands. Outside the breeding season kāhu often form communal roosts, which can contain several hundred birds. Carrion forms a major part of their diet, but they also hunt insects, small birds and mammals.

Status in Aotearoa: Native

Conservation status: Not threatened

RURU
MOREPORK

Ninox novaeseelandiae

The ruru is a well-known small (29cm long; 175g), long-tailed owl that lives in native and exotic forests throughout Aotearoa. Its distinctive *more-pork* call is commonly heard at night in forest, other rural areas and some larger urban parks. The striking yellow-green eyes are set into two facial discs either side of a small, sharply hooked bill. Birds are well camouflaged, their back feathers dark brown spotted with off-white. The dark brown breast feathers are variably streaked with cream and brown through to rust, and the legs are feathered down to their yellow feet. The short, serrated wings allow the bird to manoeuvre silently in flight through dense forest. Nocturnal, ruru are most often seen at night catching flying insects near artificial lights. In addition to their characteristic cry, they utter a repetitive *quork-quor*k, a rising *quee* and a short *yelp*.

Habitat: Roosts in dark forested areas with high overhead cover and where suitable patches of vegetation remain, including farmland, urban parks and suburbs.

Ornithologist's notes: Closely related to the Australian boobook (*Ninox boobook*) and the only one of thirty-five *Ninox* hawk-owl species to breed in Aotearoa. It is also a relative of the now extinct whēkau (laughing owl, *N. albifacies*). Monogamous pairs breed in spring and summer, nesting in holes in live or dead trees, or within crevices, burrows or large perching plants such as epiphytes. Females lay 1–3 large roundish white eggs and incubate them for about twenty-five days. Chicks fledge when they are about seven weeks old, and both adults feed the young. Ruru eat insects, moths, wētā, small rodents, small bats and small birds such as tauhou (page 133).

Status in Aotearoa: Native

Conservation status: Not threatened

KĀREAREA
NEW ZEALAND FALCON

Falco novaeseelandiae

This bold, medium-sized (40–50cm long; 200–740g) bird of prey has shorter, broader wings and a longer tail than falcon species overseas, giving it greater manoeuvrability when pursuing forest birds. Adults have a brown back, streaked cream breast, rufous undertail and 'trousers', long yellow legs, a yellow eye-ring and cere, and a facial 'moustache' stripe. Juveniles are dark brown with blue-grey legs, eye-ring and cere. Adults make a loud *kek-kek-kek* call to defend their breeding territory. Adult females and juveniles also 'whine' for food.

Habitat: Breeds from the coast to above the treeline in native forest, forest remnants, tussock grassland, roughly grazed hill country and pine forest. Widely distributed on both main islands in suitable habitat except for Te Tai Tokerau Northland, and occurring in low numbers on Rakiura Stewart Island and the subantarctic Motu Maha Auckland Islands. A few pairs have bred in Te Whanganui-a-Tara Wellington in recent years.

Ornithologist's notes: The only endemic raptor species in Aotearoa. There are two subspecies, one North Island and a larger South Island one. A recent DNA study found that the species' closest living relative is the aplomado falcon (*Falco femoralis*) of South America. Monogamous pairs nest in spring and summer in epiphytes growing on trees, or on the ground under small rocky outcrops or fallen trees. They will attack intruders, including humans, dive-bombing and striking the head with razor sharp talons. Parents share incubation of the 1–4 large brownish eggs. The female raises the chicks, while the male does most of the hunting. Kārearea eat small to medium-sized birds, and sometimes shags, poultry, common pheasants, rabbits or small hares. Juveniles may feed on cicadas and beetles when learning to hunt.

Status in Aotearoa: Endemic

Conservation status: Recovering

KŌTARE
NEW ZEALAND KINGFISHER

Todiramphus sanctus vagans

The kōtare is a beautiful medium-sized (23cm long; 55g) forest kingfisher with a bright azure-blue back and cap, and a heavy, flattened black-and-pink bill. Females are greener than males and duller above. Both have creamy-white to pale apricot undersides, broad black eye-stripes and a white collar. They can dive into water to a depth of 1m to catch prey, which they take back to their perch and eat whole. Kōtare have been observed flying aggressively at ruru (page 89) and pūkeko (page 41), including one bird that fatally speared a ruru in the eye with its sharp bill. They have a wide range of calls, the most distinctive being the staccato *kek-kek-kek* territorial call.

Habitat: Forest, mangrove, wetland, coastal and urban habitats from Te Rerenga Wairua Cape Reinga and Rangitāhua Kermadec Islands to Rakiura Stewart Island, although more common in the north. Favours river margins, farmland, urban parks, lakes, estuaries and rocky coastlines, and native forest – anywhere where there is water or open country with adjacent perches.

Ornithologist's notes: The only species of the thirty-five *Todiramphus* forest kingfishers that breeds in Aotearoa. Monogamous pairs start mating in September, followed by nest-building in October. They nest in cliffs, clay banks or tree holes. The nest chamber is made by repeatedly flying at the chosen site using the bill to chisel out dirt, then pecking out the nesting tunnel and nest chamber. Males defend a territory and females lay 5–7 small white eggs. After three weeks of incubation, mainly by the female, the chicks are fed by both parents and fledge a week later. Both parents feed them for 7–10 days after fledging, by which time they can catch their own food. Kōtare eat small crabs, tadpoles, freshwater crayfish, small fish, insects such as cicadas and stick insects, wētā, skinks, mice and small birds.

Status in Aotearoa: Native

Conservation status: Not threatened

KĀKĀPŌ

Strigops habroptila

This large (58–64cm long; 1–4kg), nocturnal, flightless parrot was once common throughout New Zealand's native forests. Predation by introduced mammals brought it to the brink of extinction, and it now exists only on predator-free islands, where intensive intervention has steadily increased numbers; 202 were alive in 2022. Kākāpō are in the same Aotearoa parrot superfamily, Strigopoidea, as kākā (page 97) and kea (page 99). They have pale owl-like faces with whiskers, and their plumage is moss green, mottled with yellow and black above, and with more yellow below. Their big bill, legs and feet are grey. The birds often leap from high trees, flapping their rather weak wings in what is at best a controlled plummet. Males make a deep booming call and a loud wheezing call to attract a mate. Both sexes make a loud, high-pitched *skraak* call.

Habitat: Nests in native forest on or under the ground in natural cavities or under dense vegetation. Forages on the ground and climbs high into trees.

Ornithologist's notes: The world's heaviest parrot, kākāpō breed in summer and autumn in years of forest fruit abundance. On islands in southern Aotearoa they breed when rimu are in fruit – usually once every 2–4 years. During the breeding season males boom from track-and-bowl systems to attract females. Most parrot species are monogamous, but kākāpō have an unusual lek-mating system in which males gather in a display area, or lek, to call and advertise their presence, and females visit the lek to choose a male to mate with. After mating, males play no part in incubation or chick-rearing. The female lays 1–4 large white eggs in a shallow depression in the soil or rotten wood and incubates them. Kākāpō are vegetarian, eating leaves, buds, flowers, fern fronds, bulbs, fruits and seeds.

Status in Aotearoa: Endemic

Conservation status: Nationally critical

KĀKĀ

Nestor meridionalis

The kākā is one of three living species in the Aotearoa parrot superfamily Strigopoidea. The related Norfolk kākā (*Nestor productus*) and Chatham Island kākā (*N. chathamensis*) are now extinct. Often heard before it is seen, the kākā is a large (38–44cm long; 340–400g) forest parrot found on all three main islands of Aotearoa and several offshore islands. It is predominantly olive-brown, with a grey-white crown, red-orange underwings, and a deep crimson belly and undertail. Males have a longer upper mandible and larger head than females. Birds make a harsh, repeated, rhythmic *ka-aa* call when flying above forest; a harsh, grating *kraak* alarm call when disturbed; and a variety of loud musical whistles.

Habitat: Inhabits a wide variety of native forest types, including podocarp, beech and pōhutukawa forest, and some urban parks in Te Whanganui-a-Tara Wellington. Now much reduced in range and abundance in both main islands due to forest loss and predation by introduced mammals, kākā are most abundant on offshore islands that have no introduced mammals, especially stoats. They have become a fairly common sight in Te Whanganui-a-Tara parks, breeding now in Karori and Ōtari-Wilton's Bush following captive releases in the 2000s. A recent study tracked a kākā making a 1000km round trip that included parts of Waikato, Te Tara-o-te-ika-a-Māui Coromandel and islands in Tīkapa Moana Hauraki Gulf.

Ornithologist's notes: Monogamous pairs mainly breed in spring and summer, usually nesting in tree cavities more than 5m above ground, or at ground level on offshore islands. The nest is lined with small wood chips. The female incubates four large white eggs and cares for the chicks, and is fed by the male through the breeding season. Both parents feed the fledglings once they get older. Kākā eat forest seeds and fruit, nectar, sap, honeydew and huhu grubs. Seasonal specialists, their diet varies as different food sources come into season.

Status in Aotearoa: Endemic

Conservation status: Recovering

KEA

Nestor notabilis

Named by Māori after its loud call, the kea is one of the smartest bird species in the world. A recent study of captive kea found that the birds could make predictions about uncertain events, behaving similarly to humans when faced with statistical reasoning tasks. The study's results mirrored those of human infants and chimpanzees in similar tests. Kea are large (46cm) green-and-olive parrots with green-blue iridescence on some feathers, and scarlet underwings. They can use their pointed grey-black bill as a third 'foot' to help them scramble over rough ground. Males are 20 percent bulkier than females, weighing 900–1100g, and have a longer bill. Juveniles have yellow skin around the bill and eyes, and a yellow lower mandible. The main call is a loud, high-pitched cry, *kee-ee-aa-aa* or *keeeeeaaaa*. Juveniles make loud whooping or squealing calls.

Habitat: Kā Tiritiri o te Moana Southern Alps and the Kaikōura ranges, where they are commonest in montane forests and adjacent subalpine–alpine zones.

Ornithologist's notes: Innately curious, and sometimes known as 'clowns of the alps', kea are attracted to Te Waipounamu South Island skifields and mountain tramping huts. Early Pākehā in the high country observed small numbers landing on sheep in winter and using their sharp bills to feed on fat deposits. This led to a mistaken belief that all kea were 'sheep killers' and a bounty was placed on the birds in 1860; more than 150,000 were killed before the species became fully protected in 1986. Monogamous pairs nest below the treeline on the ground in natural cavities and crevices. Females lay up to four large white eggs in August–October and incubate them for 22–24 days. The male brings food to her and she feeds the chicks in the nest for three months before they fledge. Kea are omnivorous, eating shoots, roots, fruits, nectar, seeds, huhu grubs and carrion. A few prey on Hutton's shearwater (*Puffinus huttoni*) nestlings.

Status in Aotearoa: Endemic

Conservation status: Nationally endangered

KĀKĀRIKI
RED-CROWNED PARAKEET

Cyanoramphus novaezelandiae

These dainty emerald-green parrots have a bright red crown, a long tail, and broad rounded wings with bright blue leading edges to their flight feathers. Males are larger than females (25–28cm long; 70–80g). The flight seems erratic when the birds are crossing open spaces, but they can fly over open ocean in excess of 100km. In flight and when alarmed, they make a prolonged squeaky chattering *kek-kek-kek* call.

Habitat: From tall native forests to grass and shrublands. Although widely distributed throughout the Aotearoa region, from subtropical Rangitāhua Kermadec Islands to the subantarctic Motu Maha Auckland Islands and Rēkohu Chatham Islands, kākāriki are almost entirely absent from the two main islands. Te Whanganui-a-Tara Wellington is an exception, where captive-bred birds have become established in Karori, Otari-Wilton's Bush, and Matiu Somes Island, and can be seen in some gardens and urban parks such as Wellington Botanic Garden ki Paekākā.

Ornithologist's notes: One of six *Cyanoramphus* parakeet species that breed in Aotearoa, with two close relatives overseas (the Norfolk parakeet, *C. cookii*, and New Caledonian parakeet, *C. saisseti*). The Māori name is used as a generic name for parakeets. Monogamous pairs have been recorded breeding year-round, with most egg-laying occurring during October–November. They nest in tree cavities, holes in the ground and on cliff faces, and under dense vegetation. The female lays 4–9 small white eggs, and is responsible for all nest preparation, incubation, brooding and feeding of the chicks until they are 10–14 days old. All food is provided to her by her mate, then both adults feed the chicks until they fledge. Kākāriki are omnivorous, eating mainly seeds, flowers and fruits, and small insects. Birds on Macauley Island in Rangitāhua have been observed scavenging shellfish.

Status in Aotearoa: Endemic

Conservation status: Relict

TĪTITIPOUNAMU
RIFLEMAN

Acanthisitta chloris

Measuring just 6g in weight and 7–9cm in length, the tiny tītitipounamu is New Zealand's smallest bird species. Along with the pīwauwau (page 105), it is one of only two surviving species of New Zealand's ancient endemic wren suborder, Acanthisitti, an early sister group to all songbirds that became isolated in Aotearoa. Very short-tailed 'fluffballs' with short wings, a slender dark bill, and large legs and feet, the birds constantly flick their wings as they forage up and down trees feeding on small insects and moths. Males are smaller than females, with a bright green cap and back. Females are mainly yellow-brown with dark speckles. Both frequently utter a high-pitched *zip*, *pip* or *chuck*.

Habitat: Mainly mature high-altitude forest, especially beech, kauri, kāmahi and podocarp. Widespread throughout Te Waipounamu South Island, but its Te Ika-a-Māui North Island distribution is patchy and includes the eastern ranges south to Remutaka, the Central Plateau, and Hauturu-o-Toi Little Barrier Island.

Ornithologist's notes: Monogamous pairs breed from August to February in cooperative family groups, in which related offspring of both sexes help to raise siblings. Unrelated helpers that assist with breeding are thought to gain pairing opportunities as a result. The birds build enclosed spherical nests within existing tree holes, dense vegetation or rock cavities. Pairs raise up to two broods per season. The male does most nest-building, with incubation and nestling and fledgling care shared. Incubation of the 2–5 small white eggs takes about twenty days, and chicks leave the nest when about twenty-four days old. The species has more than thirty Māori names, and the English name comes from a colonial-era military regiment whose uniform was a similar colour to the bird's green plumage. They eat beetles, spiders, caterpillars and moths.

Status in Aotearoa: Endemic

Conservation status: Not threatened

PĪWAUWAU
NEW ZEALAND ROCK WREN

Acanthisitta gilviventris

The pīwauwau is a small (10cm long; 16–20g), rather 'hobbit-like' bird, with long legs and very large feet. It is the only truly alpine bird species in Aotearoa, and along with tītitipounamu (page 103) is the only surviving member of New Zealand's ancient endemic wren suborder Acanthisitti. Pīwauwau have very short tails and small, rounded wings. Both sexes have pale eyebrow stripes and a short dark bill. The male is pale green above and grey-brown below with yellow flanks, and the female is mainly olive-brown. The birds spend most of their time rapidly hopping and flitting through the alpine boulder fields they inhabit, calling sporadically. When still, they bob up and down vigorously, often flicking their wings. They forage in rock crevices, under rock jumbles and in low subalpine scrub. Their flights are short and close to the ground. They make a very high-pitched three-note call and a 'whirring' call. Pairs sometimes duet.

Habitat: Alpine and subalpine areas of Kā Tiritiri o te Moana Southern Alps at 900–2500m altitude, in dense subalpine scrub, scree where stable rock falls are interspersed with low shrubbery, and bare rock in exposed situations. The birds overwinter in the alpine zone, including living in the layer under the snow.

Ornithologist's notes: Monogamous and territorial, pīwauwau breed in spring and summer. Both adults build an enclosed feather-lined nest of tussocks and grasses with a small entrance hole at ground level in a natural cavity. Females lay 3–5 small white eggs. Both adults incubate the eggs for 18–22 days and care for the young for 21–26 days until fledging, then continue feeding them for a further 2–4 weeks. Unlike tītitipounamu, pīwauwau have no nest helpers. They eat spiders, moths, wētā, beetles, nectar and small fruits.

Status in Aotearoa: Endemic

Conservation status: Nationally endangered

KORIMAKO
BELLBIRD

Anthornis melanura

The korimako is a medium-sized (20cm long; 26–34g) yellowish-green honeyeater. It has a short, decurved black bill, forked tail, red eyes and grey legs. Adult males are olive-green with paler underparts, a purple-tinted head, and blackish wings and tail. Adult females are browner, with a narrow white-yellow stripe across the cheek from the base of the bill and a bluish gloss on top of the head. The song is made up of musical ringing notes, similar to that of a tūī (page 109) but without the clicks, grunts and wheezes. In predator-free areas the population can become very high, making the chorus at dawn and dusk impressive. Joseph Banks, naturalist on HMS *Endeavour*, described a dawn chorus of korimako in Tōtaranui Queen Charlotte Sound in 1770 as 'almost imitating small bells'. The larger, darker Chatham Island bellbird (*Anthornis melanocephala*) became extinct in the early 1900s, probably due to predation by feral cats and rats.

Habitat: Native and exotic forest, scrub, urban parks and gardens throughout Te Waipounamu South Island and much of Te Ika-a-Māui North Island, although rare from Waikato northwards, except for Te Tara-o-te-ika-a-Māui Coromandel Peninsula and some offshore islands.

Ornithologist's notes: Important pollinators of flowering plants, korimako disperse the seeds of trees with small to medium-sized fruits. The species has more than twenty Māori names. Monogamous pairs breed in spring and summer, building loose nests of twigs and fibres lined with feathers and fine grasses and situated low down under dense cover. The female incubates the 3–5 small eggs, which are creamy white with brown blotches, and both parents care for the chicks. Birds are territorial when breeding but become more mobile after breeding. They have a brush-like tongue to reach inside flowers for nectar, but also feed on fruit, insects and spiders.

Status in Aotearoa: Endemic

Conservation status: Not threatened

TŪĪ

Prosthemadera novaeseelandiae

The larger (30cm long; 90–125g) of New Zealand's two honeyeater species, the tūī is a widespread and abundant songbird in the three main islands. There is also a larger Chatham Islands subspecies that is up to a third heavier. One of our most familiar native birds, the tūī has a decurved black bill, dark eyes, black legs and feet, and a black head. Its wings, tail and underparts have an iridescent blue-green or bronze sheen. The nape and side of the neck have fine white feathers, and the birds are unique in having two curved white poi, or feather tufts, on their throat. The sexes are alike, but the male is larger. Tūī are usually very vocal, with a mix of tuneful notes, coughs, grunts and wheezes, and have local dialects. Their noisy whirring flight is interspersed with short glides. In display flight, they soar upwards above the canopy and then make a near-vertical dive back down. The old English name is 'parsonbird'.

Habitat: Native forest and scrub (sometimes in exotic forests), and in rural gardens, stands of flowering kōwhai and gums, and in suburban parks and gardens.

Ornithologist's notes: Tūī play an important role in native forest ecosystems as one of the main pollinators of flowering native plants, and also disperse the seeds of trees with medium-sized fruits. The species has more than twenty Māori names. They are monogamous and notoriously aggressive defenders of their territory. Females lay 2–4 white or pale pink eggs with reddish-brown spots and blotches from September to January in a rough, bulky nest made of twigs and sticks and lined with fine grasses, high in the canopy or subcanopy. The eggs are incubated by the female and she initially feeds the chicks, with the male helping as they grow. Tūī will commute more than 10km daily to feed on rich sources of nectar and fruit. They also eat large invertebrates such as cicadas and stick insects.

Status in Aotearoa: Endemic

Conservation status: Not threatened

RIRORIRO
GREY WARBLER

Gerygone igata

This lively little warbler is New Zealand's most widely distributed endemic bird species. A small (11cm long; 5.5–6.5g) grey songbird, it vies with the slightly shorter tītitipounamu (page 103) for the title of New Zealand's smallest bird. The riroriro is olive-grey above, with a pale grey face and pale greyish-white underparts. Its tail is dark grey, with white tips that create an obvious white band in flight. The short black bill is finely pointed, the eyes bright red and the slender legs black. Birds are often seen flying short distances, moving between branches in the canopy. They have a very fast wingbeat, hovering briefly at times while foraging for insects. More often heard than seen, the male has a melancholy trilling song. Females give short *chirp* contact calls when near their male mates.

Habitat: Mid- to high levels of the forest canopy on mainland Aotearoa and most offshore islands.

Ornithologist's notes: The riroriro is most closely related to the Chatham Island warbler (*Gerygone albofrontata*). The species has more than twenty Māori names. Monogamous pairs breed in spring and summer, from July to February. The nest is a hanging enclosed dome, usually in the outer branches of the canopy a few metres off the ground. Birds in the north appear to raise one brood per season, while those in the south are typically double-brooded. Females lay 3–5 eggs per clutch. The riroriro is the only mainland host for the pīpīwharauroa (page 33), which replaces a single egg from the warbler's clutch with one of its own eggs. After hatching, the cuckoo chick ejects all eggs and/or nestlings from the nest and is raised alone by its foster parents. They are insectivorous, foraging for caterpillars, flies, beetles and moths in the bark of tree trunks and branches.

Status in Aotearoa: Endemic

Conservation status: Not threatened

KŌKAKO
NORTH ISLAND KŌKAKO

Callaeas wilsoni

With their beautiful, haunting song, kōkako evoke the forests of ancient Aotearoa. Along with the tīeke (page 115) and extinct huia (*Heteralocha acutirostris*), they are members of the Aotearoa wattlebird family, Callaeidae. Genetic studies have revealed that they share a common ancestor with the satinbirds and berrypeckers of New Guinea. The closely related orange-wattled South Island kōkako (*Callaeas cinererea*) is probably extinct. These large (38cm long; 225g) blue-grey songbirds have a long downcurved tail, a black mask around their large round black eyes, a heavy black bill, and cobalt-blue wattles that grow from the base of the bill. They use their long, strong legs to bound, hop and run among branches and on the forest floor, interspersed with glides on their short, rounded wings. The sexes are alike. Juveniles have pink or lilac wattles. Sung at dawn, their complex, melancholy song can carry for kilometres and is the longest known duet performance of any songbird.

Habitat: Typically inhabits tall native forest dominated by tawa. The species has persisted in the wild in small populations, mainly in central Te Ika-a-Māui North Island, and survives only with sustained control of introduced mammal predators.

Ornithologist's notes: Pairs are monogamous and have permanent territories. The male performs an 'archangel' courtship display to the female, raising and spreading his wings and bowing his head. Pairs typically raise one brood during November–February. The female lays 2–4 pale eggs in a cup nest in a tree, up to 25m above the ground. She incubates them for eighteen days, then both adults feed the chicks, which fledge at 32–37 days. Juveniles usually remain in the territory for a few months or up to a year, and continue to be fed by both parents. Kōkako eat fruit, leaves, flowers, nectar and some invertebrates.

Status in Aotearoa: Endemic

Conservation status: Nationally increasing

TĪEKE
NORTH ISLAND SADDLEBACK

Philesturnus rufusater

Tīeke are highly active foragers with striking black plumage, a vivid chestnut saddle across their back and bright red wattles. Medium-sized songbirds (25cm long; 70g), they are extremely vulnerable to predation by introduced mammals and can survive only at managed or predator-free sites. Tīeke are relatively weak flyers but have stout legs for bounding through forest undergrowth. The sexes are alike, although males usually have larger wattles than females of the same age. Their explosive chattering call, *cheet te-te-te-te*, resounds through the day. Territorial male birds also sing a rhythmical song of repeated phrases, over 200 variations of which have been recorded.

Habitat: Coastal and inland forests of Te Ika-a-Māui North Island. The single remaining natural population of around 500 birds is on Taranga Island off the east coast of Te Tai Tokerau Northland, but translocated populations are thriving on a number of predator-free islands, including Tiritiri Matangi and Kapiti, and at six fenced mainland sanctuaries.

Ornithologist's notes: Along with the South Island saddleback (*Philesturnus carunculatus*), kōkako (page 113), and extinct huia (*Heteralocha acutirostris*), this species is a member of the Aotearoa wattlebird family, Callaeidae. The name tīeke is also used for the South Island saddleback. Monogamous pairs typically breed in spring and summer. Females build their nests in tree cavities, rock crevices, tree-fern crowns and dense ground-level vegetation. The female lays 1–4 creamy eggs with brown speckling and incubates them, but both parents feed and care for the chicks. Tīeke eat small invertebrates, fruits and nectar. Their habit of excavating decaying or rotten wood for wood-boring grubs is woodpecker-like.

Status in Aotearoa: Endemic

Conservation status: Relict

HIHI
STITCHBIRD

Notiomystis cincta

These medium-sized (18cm long; 30–36g) songbirds are recognisable by their cocked tail, bold white wing-bar and fast flight movements in mid-canopy. Males have a black head with white ear tufts, a bright yellow neck and shoulder band, two white wing-bars, and a greyish-brown body. Females are greyish brown apart from the two white wing-bars. Juveniles resemble females, and the blackish bill is slender and downcurved. Male hihi make many variants of a loud, slurring whistle call with 2–3 notes; females may include some of these sounds in their warble. Both sexes produce a single-note *titch* warning call, along with various other whistles and warbles.

Habitat: Mature native forest in Te Ika-a-Māui North Island. Until the 1990s, few people had the opportunity to see the single remnant population of hihi on Te Hauturu-o-Toi Little Barrier Island, but thanks to successful conservation management and research, the birds can now be seen at Kapiti Island, Tiritiri Matangi, Zealandia, Maungatautari, Tarapuruhi Bushy Park and Rotokare, where they have been translocated.

Ornithologist's notes: Most closely related to the Aotearoa wattlebirds, the hihi is in its own endemic family, Notiomystidae. The species has more than twenty Māori names; hihi means 'ray of sunlight'. Males are able to raise their white ear tufts during aggressive territorial displays. Often curious, hihi will approach people for close examination while making warning calls. The birds breed in spring and summer, building a deep woven cup lined with tree-fern scales and feathers on top of a stick base inside a natural tree cavity or nest box. Females lay up to four clutches of 1–5 white eggs per season. They incubate these alone for 14–20 days, but males assist with chick-rearing, which lasts 21–31 days to fledging. Hihi eat small invertebrates, fruits and nectar.

Status in Aotearoa: Endemic

Conservation status: Nationally vulnerable

PŌPOKATEA
WHITEHEAD

Mohoua albicilla

Pōpokatea are gregarious sparrow-sized (15cm long; 15–18g) forest songsters that live in noisy groups of up to eight members. They have a compact body, a short tail and bill, and long legs. The head and underparts are white or whitish. The upperparts, wings and tail are brown-grey, and the bill, legs and eyes are dark. Reluctant flyers, they move quickly through the canopy and are acrobatic foragers, often hanging upside down anchored by their strong feet. The male makes a melodious series of canary-like whistles and trills, *viu viu viu zir zir zir zir*. Birds make several other chirps and squeaks to communicate within the group.

Habitat: Mostly the canopy regions of tall, dense forest. The species is found only in Te Ika-a-Māui North Island, south of a line connecting the Pirongia Forest, Kirikiriroa Hamilton and Te Aroha.

Ornithologist's notes: One of three species in the endemic Mohuidae family (along with mohoua, page 121, and pīpihi), it has more than thirty Māori names. The old English name is 'bush canary'. Pōpokatea form cooperative breeding groups to produce a single clutch per season from October to January. The nests are neat, tightly woven cups lined with fine materials such as natural fibres or moss, usually built in a tree fork and hidden in dense canopy vegetation. Clutches usually contain three eggs, and the female is the sole incubator. The primary male or helpers may feed her as well as her young, both on and off the nest. Pōpokatea are host to the brood-parasitic koekoeā (page 35). The cuckoo lays a single egg in the host's nest, which the pōpokatea incubates along with its own eggs. After the cuckoo chick hatches, it ejects all the eggs and/or chicks from the nest and is raised on its own. Pōpokatea eat insects such as caterpillars and beetles, spiders, and sometimes fruit or other plant material.

Status in Aotearoa: Endemic

Conservation status: Not threatened

MOHOUA
YELLOWHEAD

Mohoua ochrocephala

Mohoua are sparrow-sized (15cm long; 26–30g), yellow-headed forest songbirds. The birds have a conspicuous yellow head and breast; a brown back, wings and tail; and a white lower belly and vent. The end of the tail is tattered and spiky when worn, and the eyes, bill and legs are black. Agile when flitting from branch to branch, mohoua are surprisingly weak flyers over more than a few metres. Males make a loud, melodious series of canary-like whistles and trills, *viu viu viu zir zir zir zir*. Females make a distinctive descending buzzing call.

Habitat: Favours red beech and silver beech forest. Once one of the commonest birds of Te Waipounamu South Island and Rakiura Stewart Island forests, they have now been lost from 95 percent of their original range. A few hotspots remain in Tioripatea Haast Pass, Tāhuna Glenorchy, Te Pitonga o Te Waipounamu the Catlins and Ata Whenua Fiordland, and there are translocated populations on Whenua Hou Codfish Island, Te Wharawhara Ulva Island, and Ōruawairua Blumine Island.

Ornithologist's notes: Mohoua breed in spring and summer, mostly as monogamous pairs but with helpers at some nests. A woven, feather-lined cup of fibrous material is built in a cavity in a tree, where the female lays 1–4 small pale eggs and incubates them. Both sexes feed the young. Like pōpokatea (page 119), mohoua are the preferred hosts of the brood-parasitic koekoeā (page 35). Mohoua spend most of their time high in trees gleaning invertebrates such as caterpillars and spiders, and sometimes also eat small fruits. The old English name is 'bush canary'.

Status in Aotearoa: Endemic

Conservation status: Declining

PĪWAKAWAKA
NEW ZEALAND FANTAIL

Rhipidura fuliginosa

The pīwakawaka (16cm long; 8g) is undoubtedly one of New Zealand's best-known native birds, with its distinctive fanned tail, acrobatic flycatching and confident curiosity. There are two colour morphs, with the more common pied morph occurring throughout the species' range, and the scarce black morph comprising up to 5 percent of the Te Waipounamu South Island population but rarely found in Te Ika-a-Māui North Island. Adult pieds have a greyish head, prominent white eyebrows, a brown back and rump, a cinnamon breast and belly, black and white bands across the upper breast, and a long black-and-white tail. Juvenile pieds are mainly brown and rufous. Black pīwakawaka are mainly black, with black-brown over the rump, belly and flight feathers, and sometimes a white spot behind each eye. The loud, high-pitched song is a rhythmic whistling, *tweeta-tweeta-tweeta*.

Habitat: Native and exotic forest, shrubland, scrubland, shelter belts, orchards, and leafy suburban parks and gardens.

Ornithologist's notes: A member of the monarch flycatcher family, Rhipiduridae, the species has more than thirty Māori names. Monogamous pairs breed from August to March in the North Island, and from September to January in the South Island. The delicate cup nest is made of fine materials such as mosses, dry wood fibres and dry grasses tightly woven with cobwebs. Females lay 2–5 small pale eggs. Both adults take turns to incubate them over fourteen days, then brood and feed the young for another fourteen days. Fledglings have short tails and often remain together, perched side by side. The male looks after the fledglings when the female starts building the next nest. One monitored pair reared five broods in a season, raising fifteen fledglings. Pīwakawaka eat small invertebrates such as moths, flies, beetles and spiders, and small fruits.

Status in Aotearoa: Endemic

Conservation status: Not threatened

MIROMIRO
NEW ZEALAND TOMTIT

Petroica macrocephala

This small (13cm long; 11g), lively songbird is widespread and fairly common throughout the two main islands. It has a large head and a small black tail, large round black eyes, and dark legs with orange on the feet. Adult males have a black head, upper breast and back; black wings with a white bar; and white underparts. Females are similar but with duller black areas. In Te Waipounamu South Island, Rēkohu Chatham Islands and Motu Maha Auckland Islands, males have some yellow and/or orange coloration on the lower breast and belly; both sexes have a white spot at the base of the bill, which can be enlarged during displays. The contact call is a short *seet*, *zet* or *swee*. Individual birds can be quite confident, coming within a few metres of humans.

Habitat: Native and exotic habitats from sea level to the subalpine zone.

Ornithologist's notes: There are five subspecies: North Island tomtit, South Island tomtit, Snares Island tomtit, Chatham Island tomtit and Auckland Island tomtit. The island subspecies show a striking variation in body size, being considerably larger than their mainland relatives, a tendency known as Foster's rule. Birds from the main islands weigh around 11g, while the Snares Island tomtit on Tini Heke weighs 20g. The species has more than twenty Māori names; in Te Waipounamu its name is ngirungiru. Miromiro is also used for other tomtit species. The breeding season is mainly from September to February. Monogamous pairs rear three broods in well-concealed nests in thick vegetation or shallow cavities. Most clutches consist of 2–3 pale, speckled eggs, incubated by the female while the male provides her with food. Both feed the nestlings and fledglings. Miromiro eat small invertebrates, including spiders, beetles, flies, moths and wētā, and sometimes small fruits.

Status in Aotearoa: Endemic

Conservation status: Not threatened

KAKARUWAI
SOUTH ISLAND ROBIN

Petroica australis

The kakaruwai (18cm long; 35g) is a familiar bird in Te Waipounamu South Island backcountry and can be recognised by its long legs and upright stance. The species is territorial, and males in particular inhabit the same patch of forest throughout their lives. The male has a dark grey-black head, neck, back and upper breast. The flight and tail feathers are brownish black, and the lower breast and belly are white to yellowish white with a sharp demarcation between dark and light. Females are paler grey over the upper body, and the white breast–belly area is smaller and less clearly demarcated. Juveniles are similar, often with a smaller, or no, white front. Adults can enlarge the small white spot of feathers above the bill. Males – particularly bachelors – are melodic songsters. Where robins come in regular contact with people on walking tracks they become quite confident, often approaching to within a metre, sometimes even standing on a person's boot.

Habitat: Mature native forest, scrub and exotic tree plantations in Te Waipounamu.

Ornithologist's notes: Most closely related to the toutouwai (North Island robin, *Petroica longipes*) and kakaruia (black robin, *P. traversi*). Aotearoa robins have more than forty Māori names. Monogamous pairs start nesting in July, with the last clutches laid in December. The female builds the nest while her male mate brings her food 2–3 times per hour. Only females incubate the 2–4 white eggs with brown speckles, but both adults feed the nestlings. Nestlings leave the nest at about three weeks old and are fed by their parents for another 5–6 weeks. Kakaruwai forage on the ground, eating worms, cicadas, stick insects, tree wētā, slugs and small fruits.

Status in Aotearoa: Endemic

Conservation status: Declining

MĀTĀTĀ
FERNBIRD

Poodytes punctatus

This small (18cm long; 35g), long-tailed songbird includes five distinct subspecies, on Te Ika-a-Māui North Island, Te Waipounamu South Island, Rakiura Stewart Island, Whenua Hou Codfish Island and Tini Heke Snares Islands. The Chatham Island fernbird (*Poodytes rufescens*), which was rufous above and white below, has not been seen alive since the late 1800s. Mātātā are well camouflaged and skulk through wetland vegetation. Streaked brown above and pale below, the mainland subspecies has a chestnut cap and a prominent pale eyebrow stripe. The brown tail feathers have a distinctive tattered appearance. The birds are weak flyers; they typically scramble through dense vegetation, occasionally flying short distances with their tail hanging down. More often heard than seen, their loud *u-tick* call is given as a duet.

Habitat: Patchily distributed in wetland regions throughout Aotearoa, except Wairarapa, Te Whanganui-a-Tara Wellington and Waitaha Canterbury. Common in dry shrubland near Te Rerenga Wairua Cape Reinga and Otou North Cape, tussock flats in Tongariro and Kahurangi National Parks, pākihi vegetation on Te Tai o Poutini West Coast, and saltmarsh reedbeds in Te Tai o Poutini, Ōtākou Otago and Murihiku Southland.

Ornithologist's notes: Closely related to the striated grassbird (*Megalurus palustris*) of Australia and Asia, and sometimes included in the same genus. Monogamous pairs breed in spring and summer, nesting in a deep, feather-lined cup of fine grass or leaves hidden among dense vegetation. Both members of the breeding pair incubate the 2–4 eggs, which are dull white or pinkish with violet and purple-brown speckles, mainly at the larger end. The pair continue to care for the young for fifteen days. Mātātā feed on insects such as caterpillars, flies, beetles and moths, spiders, and occasionally seeds and fruits.

Status in Aotearoa: Endemic

Conservation status: Declining

WAROU
WELCOME SWALLOW

Hirundo neoxena neoxena

Warou are small (14–16cm long; 9–20g), streamlined, fast-flying swallows of open country near water, and are highly adapted to aerial feeding on small insects. They are striking, elegant birds with rounded heads, a deeply forked tail and long, pointed wings. Adults have an orange-red forehead, face and breast, and black eye-stripes. The short bill is broad and black, and the back and upper wings are metallic blue-black. The underparts are pale buff, and the dark tail has ten white spots near the feather tips that form a row when the tail is spread in flight. The females' tail streamers are slightly shorter and the tail spots smaller. Juveniles are similar, but with a darker head and duller colouring. Warou make quiet pīwakawaka-like twittering, chattering and chirruping calls.

Habitat: Open country around lakes, coasts, riverbeds and ponds.

Ornithologist's notes: Self-introduced from Australia, warou were rare vagrants to Aotearoa before the late 1950s. The first record of breeding was at Awanui in Te Tai Tokerau Northland in 1958 and numbers then increased markedly, particularly in Te Ika-a-Māui North Island. By 1965 the birds were common throughout Te Tai Tokerau, spreading elsewhere in Te Ika-a-Māui and breeding in Te Waipounamu South Island. Warou raise up to three broods between August and February, in a distinctive cup-shaped nest built on a ledge or attached to a vertical support, usually on a built structure such as a house, bridge and culvert out of direct sunlight. The nest is built from the base upwards using mud and grasses, and then lined with fine grasses and feathers. The female lays 3–5 pinkish eggs with brown speckles and incubates them for fifteen days. The nestlings are fed by both parents and fledge at eighteen days. The nests are often reused within and between breeding seasons. Warou forage aerially for small invertebrates.

Status in Aotearoa: Native

Conservation status: Not threatened

TAUHOU
SILVEREYE

Zosterops lateralis lateralis

Tauhou are small (12cm long; 13g) songbirds, easily recognised by their conspicuous silvery-white eye-ring. They have green-olive plumage on the head, lower back, tail and wings, and grey on the upper back. The underparts are creamy grey with buff-pink across the flanks. The bill is dark, fine and short, and the legs pale brown. Tauhou produce a range of clear, melodious calls, including warbles and trills. The main contact call is a plaintive *creee*, and the flight call a shorter *cli-cli*. The birds are gregarious and well known for flocking in large groups, especially in winter.

Habitat: Most areas with vegetation, including suburban gardens, farmland, orchards, woodlands and native and exotic forests. Tauhou migrated to Aotearoa from Australia in the 1850s, and are now one of our most abundant and widespread native bird species, found on mainland, offshore and outlying islands.

Ornithologist's notes: One of 107 species in the *Zosterops* white-eye genus of Australia, New Guinea, Asia and Africa, members of which have also colonised many islands of the south-west Pacific. Monogamous pairs nest between August and February. The delicate cup nest is woven by one or both adults from moss, lichen, fine twigs, hair, spider web and thistledown into the small outermost branches of a tree, shrub or tree fern, usually more than 8m above the ground. Pairs may raise 2–3 clutches during a season, with 2–4 small, pale blue eggs per clutch. Both parents share incubation, which takes 10–12 days. The chicks are then fed by both parents and fledge after 9–11 days. Tauhou are omnivorous and eat a range of insects, fruits and nectar. They help with the pollination of some native tree species, including kōwhai and kōtukutuku (tree fuchsia), while feeding on their nectar.

Status in Aotearoa: Native

Conservation status: Not threatened

PĪHOIHOI
NEW ZEALAND PIPIT

Anthus novaeseelandiae

A member of the wagtail family, Motacillidae, the pīhoihoi is a small (18cm long; 35g), slender and approachable songbird that resembles a Eurasian skylark but with longer legs, and that walks rather than hops. Well camouflaged, pīhoihoi are mottled grey-brown, with a brown-streaked white breast, prominent pale eyebrow stripes, and white outer tail feathers. In common with wagtails, they frequently flick their long tail, which is most noticeable when they are standing still or perched. Their distinctive call, heard throughout the year, is a strident *tzweep*.

Habitat: Widespread in open country along coastlines, wetlands, rivers and farmland, as well as on scree faces, tussock grasslands and alpine shrublands. Pīhoihoi can be found throughout Aotearoa, from Te Tai Tokerau Northland to subantarctic Motu Ihupuku Campbell Island, and from sea level to the alpine zone.

Ornithologist's notes: Monogamous pairs breed from August to March. The female weaves a cup nest of grass under tussocks or grass clumps within ferns, and partially or fully covered with vegetation. The clutch size is typically 2–4 eggs, which are cream-coloured with large brown blotches. Incubation is by both adults and takes 14–16 days, and chicks fledge at fourteen days. Both adults feed the young on the nest. The number of clutches reared per year is unknown. Omnivorous, pīhoihoi eat grains, seeds and small invertebrates.

Status in Aotearoa: Endemic

Conservation status: Declining

GLOSSARY

Braying A loud, repeated sound like that of a donkey.

Brood (noun) A set of young birds, or baby bird siblings, hatched at the same time by the same parents.

Brood (verb) To cover chicks or young birds with the wings or feet.

Brood parasite A bird species that relies on another bird species to raise its young.

Cere A soft, waxy, fleshy swelling at the base of a bird's bill containing the nostrils.

Colour morph One of various distinct colour forms or variants of a species.

Creche A group of offspring in a colony, often cared for by several adults.

Cryptic Colouration or markings that camouflage a bird in its natural environment.

Decurved Curved downward.

Endemic species A species found in a single defined geographical location, such as an island or a country, in this case Aotearoa.

Filoplume Hair-like feathers with a few soft barbs near the tips that may have a sensory or decorative function.

Fledge To develop wing feathers that are large enough for flight.

Krill A small shrimp-like planktonic crustacean of the open seas.

Lek A communal area where two or more males of a species perform courtship displays.

Moult The process of shedding and regrowing feathers.

Pākihi A type of wet heath habitat characterised by very infertile soils.

Rail Member of a large, widespread family of more than 130 ground-living bird species that occur on all continents except Antarctica, almost all islands, and all habitats except desert.

Raptor A bird of prey that feeds on live captured prey or carrion.

Relict A population or species of bird that was once more widespread or diverse.

Spacing calls Bird calls that advertise the presence of a territory to other birds of the same species.

Subspecies One of two or more populations of a species living in different subdivisions of the species' range and varying from one another by morphological characteristics.

Translocation The managed movement of live birds or plants from one location to another.

Vagrant A bird outside its usual breeding and wintering range.

Vestigial wings Remnants of a flying ancestors' wings.

Wattles Paired fleshy growths hanging from the head or face of a bird.

REFERENCES

Checklist Committee (OSNZ), *Checklist of the Birds of New Zealand*, 5th edition, Ornithological Society of New Zealand Occasional Publication No.1, Ornithological Society of New Zealand, Wellington, 2022.

New Zealand Birds Online – The digital encyclopaedia of New Zealand birds, www.nzbirdsonline.org.nz, accessed 9 June 2022.

Tennyson, A. and P. Martinson, *Extinct Birds of New Zealand*, Te Papa Press, Wellington, 2007.

ACKNOWLEDGEMENTS

We would like to thank Colin Miskelly, Curator of Vertebrates at Te Papa and editor of *New Zealand Birds Online*, for allowing us to use the website as the primary source for our text.

Many of the illustrations in this book first appeared in *Native Birds* and *More Birds* by Charles Masefield, published by AH & AW Reed in 1948 and 1951 respectively. Penguin Random House New Zealand now holds the rights for those books and has kindly given permission to reproduce them here. Where there was no original artwork for a species, Pippa Keel has produced beautiful new illustrations and, where required, made adjustments to a few of the originals.

Thanks also to Tim Denee for the book and series design, Susi Bailey for the copy edit, and Teresa McIntyre for the proof read.

INDEX OF SPECIES

Bold page numbers refer to species descriptions.

ABOUT THE AUTHORS

Michael Szabo is editor of *Birds New Zealand* magazine and a significant contributor to *New Zealand Birds Online*. He was principal author of *Wild Encounters: A Forest & Bird guide to discovering New Zealand's unique wildlife* (2009), a former editor of *Forest & Bird* magazine, and has written for *New Scientist, New Zealand Geographic* and the *Sunday Star-Times*.

Alan Tennyson is a Curator of Vertebrates at Te Papa. His main research interests relate to the biogeography of New Zealand's biota, extinction, bird palaeontology, bird taxonomy, and population monitoring and conservation of seabirds. He is also a contributor to *New Zealand Birds Online* and wrote *Extinct Birds of New Zealand* with Paul Martinson.

ABOUT THE ILLUSTRATOR

Pippa Keel is an award-winning illustration designer, who has an Honours degree in illustration and a huge love of the outdoors. From her small studio in Wellington, Pippa has worked with a variety of New Zealand-based companies and publishers, including Zealandia Ecosanctuary and Te Papa Press (*The Nature Activity Book*, 2020).

First published in New Zealand in 2022 by
Te Papa Press, PO Box 467, Wellington, New Zealand
www.tepapapress.co.nz

Text: Michael Szabo and Alan Tennyson (Introduction)
© Museum of New Zealand Te Papa Tongarewa

Illustrations on pages 18, 20, 22, 24, 26, 28, 34, 36, 46, 48, 50, 54, 58, 60, 62, 64, 66, 70, 74, 78, 80, 84, 92, 104, 112, 126, 130 and 132 are by Pippa Keel
© Museum of New Zealand Te Papa Tongarewa

Illustrations on pages 16, 30, 38, 40, 42, 52, 56, 68, 72, 76, 86, 88, 98, 102 and 124 are reproduced with the permission of Penguin Random House, with adjustments by Pippa Keel.

All others are reproduced with the permission of Penguin Random House.

This book is copyright. Apart from any fair dealing for the purpose of private study, research, criticism, or review, as permitted under the Copyright Act, no part of this book may be reproduced by any process, stored in a retrieval system, or transmitted in any form, without the prior permission of the Museum of New Zealand Te Papa Tongarewa.

TE PAPA® is the trademark of the Museum of New Zealand Te Papa Tongarewa
Te Papa Press is an imprint of the Museum of New Zealand Te Papa Tongarewa

A catalogue record is available from the National Library of New Zealand

ISBN 978-1-99-115094-3

Cover and internal design by Tim Denee
Cover illustrations based on the tūī, North Island kōkako, tīeke, and hihi

Printed by Everbest Printing Investment Limited